SpringerBriefs in Applied Sciences and Technology

SpringerBriefs present concise summaries of cutting-edge research and practical applications across a wide spectrum of fields. Featuring compact volumes of 50 to 125 pages, the series covers a range of content from professional to academic.

Typical publications can be:

- A timely report of state-of-the art methods
- An introduction to or a manual for the application of mathematical or computer techniques
- A bridge between new research results, as published in journal articles
- A snapshot of a hot or emerging topic
- An in-depth case study
- A presentation of core concepts that students must understand in order to make independent contributions

SpringerBriefs are characterized by fast, global electronic dissemination, standard publishing contracts, standardized manuscript preparation and formatting guidelines, and expedited production schedules.

On the one hand, **SpringerBriefs in Applied Sciences and Technology** are devoted to the publication of fundamentals and applications within the different classical engineering disciplines as well as in interdisciplinary fields that recently emerged between these areas. On the other hand, as the boundary separating fundamental research and applied technology is more and more dissolving, this series is particularly open to trans-disciplinary topics between fundamental science and engineering.

Indexed by EI-Compendex, SCOPUS and Springerlink.

More information about this series at https://link.springer.com/bookseries/8884

Vítor João Pereira Domingues Martinho

Trends of the Agricultural Sector in Era 4.0

Vítor João Pereira Domingues Martinho
Agricultural School (ESAV)
and CERNAS-IPV Research Centre
Polytechnic Institute of Viseu (IPV)
Viseu, Portugal

ISSN 2191-530X   ISSN 2191-5318 (electronic)
SpringerBriefs in Applied Sciences and Technology
ISBN 978-3-030-98958-3   ISBN 978-3-030-98959-0 (eBook)
https://doi.org/10.1007/978-3-030-98959-0

This Springer imprint is published by the registered company Springer Nature Switzerland AG
The registered company address is: Gewerbestrasse 11, 6330 Cham, Switzerland

Acknowledgements

This work is funded by National Funds through the FCT, Foundation for Science and Technology, I.P., within the scope of the project Refª UIDB/00681/2020. Furthermore, we would like to thank the CERNAS Research Centre and the Polytechnic Institute of Viseu for their support.

I would also like to thank all those who have contributed in some way to this work.

A special thanks to my wife Lúcia Domingues Martinho and my two daughters Inês Domingues Martinho and Isabel Domingues Martinho.

About This Book

This book is an overview of the impacts on the agricultural sector worldwide of Era 4.0, highlighting dimensions related to Agriculture, Food and Industry (4.0). The main topics discussed are those associated with IoT in the various sectors of the economy and their impacts on the sustainability of the farms. Specifically, the following dimensions were addressed: impacts of Era 4.0 on agriculture around the world and on different farming activities; relations between Agriculture 4.0, Food 4.0 and Industry 4.0 and the various dimensions of Agriculture 4.0. There is a deep concern with the relationship between sustainability and agricultural competitiveness, where the smart concept is called to contribute. This book can be a relevant contribution, in these domains, with interesting insights for policymakers, students, researchers and economic stakeholders.

Contents

About the Author

Vítor João Pereira Domingues Martinho is Coordinator Professor with Habilitation at the Polytechnic Institute of Viseu, Portugal, and holds a Ph.D. in Economics from the University of Coimbra, Portugal. He was President of the Scientific Council, President of the Directive Council and President of the Agricultural Polytechnic School of Viseu, Portugal, from 2006 to 2012. He was an Erasmus student in the Faculty of Economics from the University of Verona, Italy; has participated in various technical and scientific events nationally and internationally; has published several technical and scientific papers; is a referee of some scientific and technical journals and participates in the evaluation of national and international projects.

Chapter 1
Bibliometric Analysis on Era 4.0: Main Highlights for the Agricultural Sector

1.1 Introduction

There are not many studies relating to Agriculture 4.0 and bibliometric analysis. The same occurs for the topics Food 4.0 and bibliometric. Nonetheless, for the subjects Industry 4.0 and bibliometric analysis, the findings are different and there are, in fact, many more documents. As follows, in this section, contributions from some of these studies have been highlighted.

Agriculture 4.0 is associated with the consideration of information and technologies to be incorporated into agricultural equipment for better efficiency and precision in the farming sector [1]. A more efficient and competitive agricultural sector is determinant to improve the competitiveness of agriculture and to address sustainability goals [2]. Agriculture 4.0 contexts are associated with technologies such as IoT (Internet of Things), machine learning and geostatistics [3].

Another way to deal with the challenges created by climate change and the increasing need for food may be to better understand consumers' behaviour and their attitudes towards food. This further understanding allows for better design approaches towards more sustainable food consumption [4].

For Industry 4.0, the implementation of robotics and automation are among the main tasks for the several activities developed in this sector, including the construction [5] and building industry [6]. Robotic technologies may bring relevant added value into different industrial activities, such as logistics [7]. Digital tools and smart objects are other approaches often considered by companies in the Era of Industry 4.0 [8]. The literature about these topics has increased in recent years, IoT and Cyber-Physical Systems (CPS) are some of the most considered keywords [9] and China and USA are among the leading countries in terms of research on these issues [10]. Nonetheless, other countries, such as Brazil, have also made their contributions [11]. The concepts associated with Industry 4.0 appear, indeed, as interrelated with diverse contexts and dimensions, for example, those related with the sharing economy [12], Big Data [13], Sustainable Manufacturing 4.0 [14], apparel

© The Author(s), under exclusive license to Springer Nature Switzerland AG 2022
V. J. P. D. Martinho, *Trends of the Agricultural Sector in Era 4.0*,
SpringerBriefs in Applied Sciences and Technology,
https://doi.org/10.1007/978-3-030-98959-0_1

industry [15], marketing [16], Cloud Manufacturing [17], artificial intelligence, neural networks and data mining [18] and sustainable development [19]. These multidimensional characteristics of Industry 4.0 technologies call for multidisciplinary approaches [20].

Considering these frameworks, it seems pertinent to analyse the several bibliographic interrelationships of Era 4.0 with agricultural sector dynamics. In this way, several documents, in a search carried out on 23 December 2021, from the Web of Science Core Collection (WoS) [21] and Scopus [22] databases for the topics "agricultur* 4.0", "agricultur* and food 4.0" and "agricultur* and industry 4.0" were considered. These documents were analysed through bibliometric analysis, following the VOSviewer [23, 24] software procedures, to highlight the main dimensions from the literature about the main sources, countries, organisations and authors. For the bibliometric analysis, bibliographic data for the following links were considered: co-authorship; co-occurrence; citation; bibliographic coupling; and co-citation. The term "agricultur*" was used as it allows for a broader search and captures expressions that encompass words such as "agriculture" and "agricultural" [25].

The findings which were highlighted by this research may be considered as a relevant basis for the several stakeholders associated with the different dimensions of the agricultural sector, specifically policymakers. In fact, the new challenges created by recent realities demand new approaches in the design of policy measures [26, 27].

The remaining structure of this study is organised into 3 more sections for each individual topic analysed ("agricultur* 4.0", "agricultur* and food 4.0" and "agricultur* and industry 4.0") and a final section for the conclusions.

1.2 Bibliometric Analysis for the Topic "Agricultur* 4.0"

For this topic, 122 and 218 documents were found on the WoS and Scopus databases, respectively. This section is organised into subsections for every link provided by the VOSviewer software for bibliographic data and inside each subsection, the metrics for the different items are highlighted. In this way, this section is divided into 5 subsections for each link assessed (co-authorship; co-occurrence; citation; bibliographic coupling; and co-citation). The main metrics are presented in tables and figures (network visualisation maps). In figures, each colour corresponds to a cluster.

1.2.1 Co-authorship

The relatedness for these links is based on the number of co-authored documents. One was always considered as the minimum number of documents of an item. The dimension of the circle associated with each item shows the number of documents.

From the WoS data, Fig. 1.1 and Table 1.1 shows that, for this link (co-authorship), the number of documents networked for the item authors is reduced, recent and without citations. The same happens for the item organisations (Fig. 1.3 and Table 1.3). For the item countries, the number of documents networked is higher (Fig. 1.2 and Table 1.2). Brazil is among the countries with a greater number of documents (with high relatedness with Portugal and Spain). England, Spain, Netherlands, France and Australia are the top 5 countries having more citations.

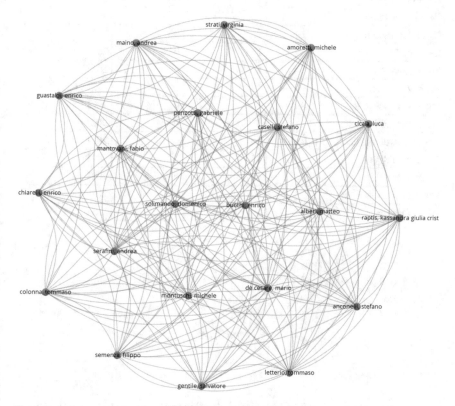

Fig. 1.1 Authors networked from WoS for co-authorship links

Table 1.1 Authors networked from WoS for co-authorship links

Authors	Documents	Citations	Avg. pub. year
Alberi, Matteo	1	0	2021
Amoretti, Michele	1	0	2021
Anconelli, Stefano	1	0	2021
Bucchi, Enrico	1	0	2021

(continued)

Table 1.1 (continued)

Authors	Documents	Citations	Avg. pub. year
Caselli, Stefano	1	0	2021
Chiarelli, Enrico	1	0	2021
Cicala, Luca	1	0	2021
Colonna, Tommaso	1	0	2021
De cesare, Mario	1	0	2021
Gentile, Salvatore	1	0	2021
Guastaldi, Enrico	1	0	2021
Letterio, Tommaso	1	0	2021
Maino, and Rea	1	0	2021
Mantovani, Fabio	1	0	2021
Montuschi, Michele	1	0	2021
Penzotti, Gabriele	1	0	2021
Raptis, Kassandra Giulia Cristina	1	0	2021
Semenza, Filippo	1	0	2021
Serafini, and Rea	1	0	2021
Solimando, Domenico	1	0	2021
Strati, Virginia	1	0	2021

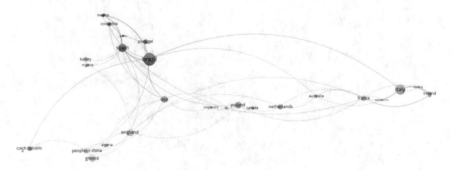

Fig. 1.2 Countries networked from WoS for co-authorship links

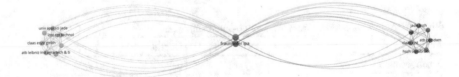

Fig. 1.3 Organisations networked from WoS for co-authorship links

Table 1.2   Countries networked from WoS for co-authorship links

Countries	Documents	Citations	Avg. pub. year
England	10	320	2020
Spain	12	266	2020
Netherlands	5	261	2020
France	5	205	2020
Australia	2	131	2020
Italy	18	101	2020
Poland	8	85	2020
Brazil	29	58	2020
Peoples r china	5	53	2021
Colombia	3	35	2020
Canada	2	34	2021
Greece	9	31	2021
South Africa	1	26	2021
Mexico	2	21	2021
Scotland	1	21	2020
Algeria	2	17	2021
USA	8	15	2020
Portugal	4	14	2021
India	2	9	2021
Ireland	3	6	2019
Romania	3	5	2020
Singapore	1	4	2021
Czech Republic	4	3	2021
Nigeria	2	1	2021
Turkey	4	1	2020
Estonia	1	0	2021
Morocco	1	0	–
Peru	1	0	2020
Russia	1	0	2021
Switzerland	1	0	2020

Table 1.3 Organisations networked from WoS for co-authorship links

Organisations	Documents	Citations	Score <Avg. pub. year>
Atb Leibniz Inst Agrartech & Biookon	1	0	2021
Atb Potsdam	1	0	2021
Claas Esyst gmbh	1	0	2021
Claas Syst	1	0	2021
Fraunhofer Ipa	2	0	2021
Hsch Osnabruck	1	0	2021
Inst Agr Technol	1	0	2021
Jade Hsch	1	0	2021
Johann Heinrich Thunen Inst	1	0	2021
Krone	1	0	2021
Maschinenfabr Bernard Krone gmbh & co kg	1	0	2021
Tech Univ Munich	2	0	2021
Univ Appl Sci Jade	1	0	2021
Univ Appl Sci Osnabrueck	1	0	2021
Vdi	1	0	2021
Vdi Assoc German Engineers	1	0	2021

The information obtained from the Scopus database for the items in this link (Figs. 1.4, 1.5 and 1.6 and Tables 1.4, 1.5 and 1.6) confirms the trends highlighted for the documents found on the WoS database. It is worth stressing that the decreasing order for the top 5 most cited countries is as follows: Netherlands, United Kingdom, France, Spain and Australia (Fig. 1.5 and Table 1.5).

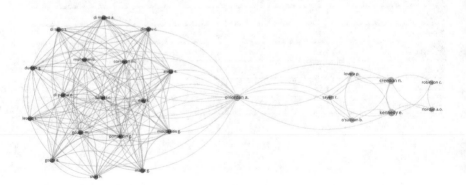

Fig. 1.4 Authors networked from Scopus for co-authorship links

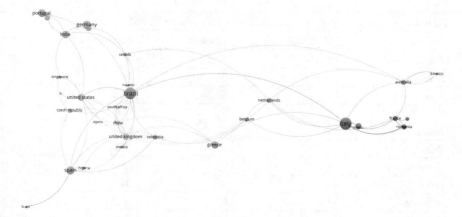

Fig. 1.5 Countries networked from Scopus for co-authorship links

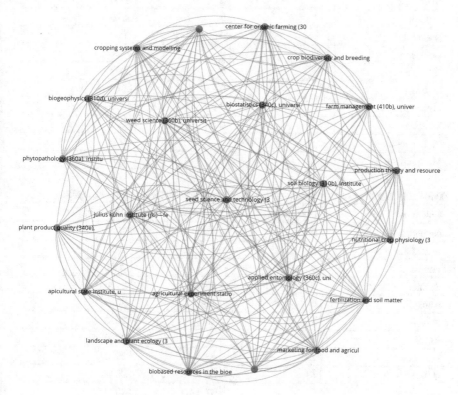

Fig. 1.6 Organisations networked from Scopus for co-authorship links

Table 1.4 Top 20 most cited authors networked from Scopus for co-authorship links

Authors	Documents	Citations	Avg. pub. year
Creedon n	2	5	2019
Kennedy e	2	5	2019
Riordan a.o	1	5	2019
Robinson c	1	5	2019
Alessi e	1	0	2020
Balan t.c	1	0	2020
Correvon m	1	0	2020
Daly k	1	0	2020
Di Matteo a	1	0	2020
Di Palma v	1	0	2020
Di Salvo s	1	0	2020
Dimitru c	1	0	2020
Dudnik g	1	0	2020
Gougis m	1	0	2020
Gouze e	1	0	2020
Lesecq s	1	0	2020
Lovera p	1	0	2019
Mailat g	1	0	2020
Molleman b	1	0	2020
Mouzakitis g	1	0	2020

Table 1.5 Top 20 most cited countries networked from Scopus for co-authorship links

Countries	Documents	Citations	Avg. pub. year
Netherlands	6	334	2020
United Kingdom	11	315	2020
France	6	260	2020
Spain	14	251	2020
Australia	6	174	2020
Italy	32	160	2020
Brazil	31	73	2020
Portugal	13	53	2020
Greece	12	52	2021
Colombia	6	47	2020
Malaysia	8	46	2020
China	6	42	2021
Germany	15	42	2019
Canada	3	39	2021

(continued)

Table 1.5 (continued)

Countries	Documents	Citations	Avg. pub. year
Hong Kong	2	31	2021
South Africa	3	31	2021
Poland	7	30	2020
Mexico	2	29	2021
United States	11	23	2020
Romania	6	22	2019

Table 1.6 Organisations networked from Scopus for co-authorship links

Organisations	Documents	Citations	Avg. pub. year
Agricultural Experiment Station, University of Hohenheim	1	0	2021
Apicultural State Institute, University of Hohenheim	1	0	2021
Applied Entomology, University of Hohenheim	1	0	2021
Biobased Resources in the Bioeconomy, Institute of Crop Science, University of Hohenheim	1	0	2021
Biogeophysics, University of Hohenheim	1	0	2021
Biostatistics, University of Hohenheim	1	0	2021
Center for Organic Farming, University of Hohenheim	1	0	2021
Crop Biodiversity and Breeding Informatics, University of Hohenheim	1	0	2021
Cropping Systems and Modelling, Institute of Crop Science, University of Hohenheim	1	0	2021
Farm Management, University of Hohenheim	1	0	2021
Fertilization and Soil Matter Dynamics, University of Hohenheim	1	0	2021
Institute of Applied Agriculture (IAAF), Nuertingen Geislingen University	1	0	2021
Julius Kühn Institute (JKI)—Federal Research Centre for Cultivated Plants, Institute for Strategies and Technology Assessment	1	0	2021
Landscape and Plant Ecology, University of Hohenheim	1	0	2021
Marketing for Food and Agricultural Products, Department of Agricultural Economics and Rural Development, University of Göttingen	1	0	2021
Nutritional Crop Physiology, Institute of Crop Science, University of Hohenheim	1	0	2021
Phytopathology, Institute of Phytomedicine, University of Hohenheim	1	0	2021
Plant Product Quality, University of Hohenheim	1	0	2021

(continued)

Table 1.6 (continued)

Organisations	Documents	Citations	Avg. pub. year
Production Theory and Resource Economics, University of Hohenheim	1	0	2021
Seed Science and Technology, University of Hohenheim	1	0	2021
Soil Biology, Institute of Soil Science and Land Evaluation, University of Hohenheim	1	0	2021
Technology in Crop Production, University of Hohenheim	1	0	2021
Weed Science, University of Hohenheim	1	0	2021

1.2.2   Co-occurrence

For these links, the relatedness is associated with the number of documents in which the items occur together. One was considered as the minimum number of occurrences of an item. The dimension of the circle associated with each item shows the number of occurrences (number of documents in which a keyword occurs).

The keywords with more occurrences, for the item all keywords, in the WoS database were the following (Fig. 1.7 and Table 1.7): agriculture 4.0; precision agri-

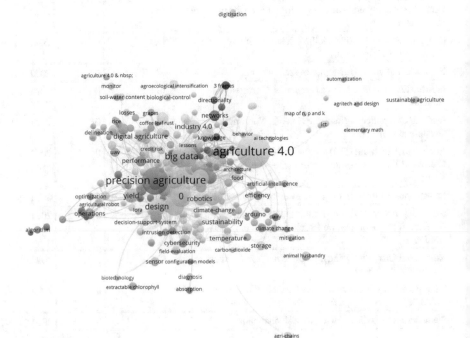

Fig. 1.7 All keywords networked from WoS for co-occurrence links

Table 1.7 Top 20 all keywords networked, with more occurrences, from WoS for co-occurrence links

All keywords	Occurrences	Avg. pub. year
Agriculture 4.0	42	2020
Precision agriculture	26	2020
Big data	16	2020
Internet of things	15	2020
System	14	2020
Agriculture 4	13	2021
Agriculture	11	2020
Design	11	2020
Management	11	2020
Smart farming	11	2020
Internet	10	2020
Technology	10	2020
IoT	9	2020
Artificial intelligence	8	2020
Systems	8	2020
Technologies	8	2020
Things	7	2021
Yield	7	2020
Challenges	6	2020
Industry 4.0	6	2020

culture; big data; internet of things; and system. For the item author keywords (Fig. 1.8 and Table 1.8), the most considered keywords were: agriculture 4.0; precision agriculture; internet of things; agriculture 4; and smart farming. From the Scopus information, among the top 5 keywords having higher occurrences, the following terms also appear (Figs. 1.9 and 1.10 and Tables 1.9 and 1.10): agricultural robots; and industry 4.0.

1.2.3 Citation

In the citation links, the relatedness is based on the number of times the items cite each other. For the minimum number of documents for the items analysed, a value of one was considered. The dimension of the label associated with each item shows the number of documents.

The findings shown in the figures and tables of this subsection confirm that the topic Agriculture 4.0 is recent. In addition, from the WoS documents, the most cited authors are the following (Fig. 1.11 and Table 1.11): Klerkx, Laurens; Jakku, Emma; Labarthe, Pierre; Rose, David Christian; and Chilvers, Jason. Figure 1.12 and Table 1.12 confirm the trends found for the most cited countries in the links co-authorships. The most cited organisations are these (Fig. 1.13 and Table 1.13): Wageningen Univ; Univ Politecn Valencia; Csiro Land and Water; Inra; and Univ Reading. The findings for the most cited sources are the following (Fig. 1.14 and Table 1.14): Njas-Wageningen Journal of Life Sciences; Agronomy-Basel; Frontiers in Sustainable Food Systems; Processes; and Computers in Industry. The results obtained with the information from the Scopus database (Figs. 1.15, 1.16, 1.17 and 1.18 and Tables 1.15, 1.16, 1.17 and 1.18) confirm, in general, these tendencies described for the documents from WoS.

1.2.4 Bibliographic Coupling

The relatedness concerns the number of references the items share. One was taken into account as being the minimum number of documents for the items analysed. The dimension of the label represents the number of documents.

The findings displayed in Figs. 1.19, 1.20, 1.21, 1.22, 1.23, 1.24, 1.25 and 1.26 and Tables 1.19, 1.20, 1.21, 1.22, 1.23, 1.24, 1.25 and 1.26 confirm the results found earlier for the several items of the link citation. This means that for the topic addressed here ("agricultur* 4.0"), the results in terms of items networked and respective citations are similar. Nonetheless, the relatedness highlighted, namely by the figures is different across links and databases. For example, for the link citation and WoS database, the country with more documents (Brazil) has a greater relatedness with England and Italy (Fig. 1.12), but for the Scopus platform, the greater relatedness

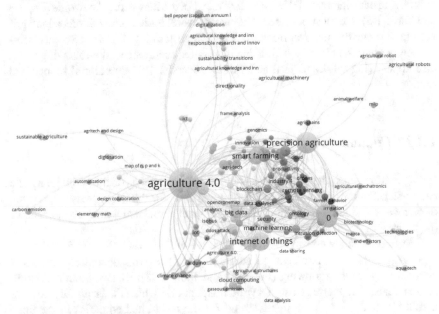

Fig. 1.8 Author keywords networked from WoS for co-occurrence links

Table 1.8 Top 20 author keywords networked, with more occurrences, from WoS for co-occurrence links

Author keywords	Occurrences	Avg. pub. year
Agriculture 4.0	42	2020
Precision agriculture	20	2020
Internet of things	15	2020
Agriculture 4	13	2021
Smart farming	11	2020
Agriculture	8	2020
Artificial intelligence	8	2020
Big data	6	2020
Industry 4.0	6	2020
Robotics	6	2021
Digital agriculture	5	2020
Machine learning	5	2020
Precision farming	5	2020
Smart agriculture	5	2021
Sustainability	5	2020
Remote sensing	4	2020
Sustainable intensification	4	2020
Arduino	3	2021
Blockchain	3	2021
Cloud computing	3	2021

of this country is with Canada, Mexico and Portugal (Fig. 1.16). For the link bibliographic coupling, the greater relatedness of Brazil for WoS documents is with Russia and Ireland (Fig. 1.20) and for Scopus studies is with Colombia, Germany, Turkey, China and India (Fig. 1.24).

1.2.5 Co-citation

For these links, the relatedness of the items is associated with the number of times they are cited together. One was considered as the minimum number of citations of an item. The dimension of the circle associated with each item shows the number of citations.

For this link and WoS studies, the most cited authors are the following (Fig. 1.27 and Table 1.27): Klerkx, L; Rose, Dc; FAO; Wolfert, S; and Liu, Y. For the Scopus documents the most cited authors are (Fig. 1.29 and Table 1.29): Klerkx, L.; Rose,

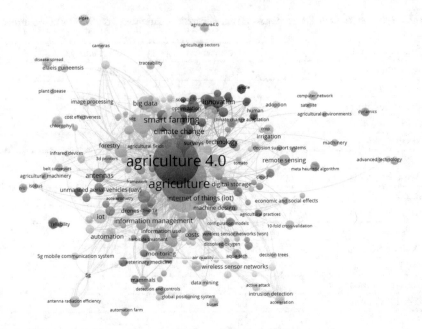

Fig. 1.9 All keywords networked from Scopus for co-occurrence links

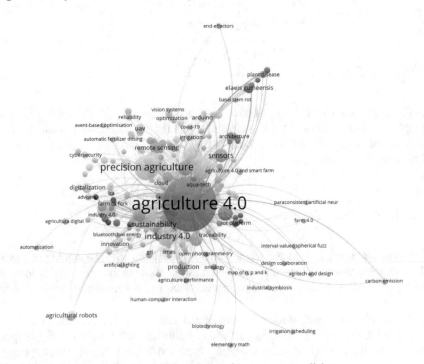

Fig. 1.10 Author keywords networked from Scopus for co-occurrence links

Table 1.9 Top 20 all keywords networked, with more occurrences, from Scopus for co-occurrence links

All keywords	Occurrences	Avg. pub. year
Agriculture 4.0	117	2020
Agriculture	70	2020
Agricultural robots	64	2021
Internet of things	46	2020
Precision agriculture	28	2020
Smart farming	24	2020
Artificial intelligence	21	2020
Industry 4.0	21	2020
Digital agriculture	18	2020
Decision making	15	2020
Climate change	14	2020
Machine learning	14	2020
Robotics	13	2020
Sustainability	13	2020
Antennas	10	2020
Big data	10	2021
Blockchain	10	2021
Crops	10	2020
Digital transformation	10	2020
Food supply	10	2021

Table 1.10 Top 20 author keywords networked, with more occurrences, from Scopus for co-occurrence links

Author keywords	Occurrences	Avg. pub. year
Agriculture 4.0	117	2020
Precision agriculture	24	2020
Smart farming	24	2020
Internet of things	22	2020
Industry 4.0	15	2020
Artificial intelligence	14	2020
Digital agriculture	14	2020
Machine learning	13	2020
IoT	10	2021
Smart agriculture	9	2021
Sustainability	9	2020

(continued)

Table 1.10 (continued)

Author keywords	Occurrences	Avg. pub. year
Agriculture	8	2020
Big data	8	2020
Blockchain	8	2021
Robotics	8	2021
Precision farming	7	2020
Sensors	7	2021
Digital transformation	6	2021
Remote sensing	5	2020
Technology	5	2020

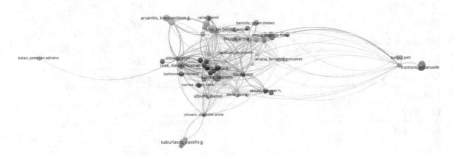

Fig. 1.11 Authors networked from WoS for citation links

D.C.; Jakku, E.; Wolfert, S.; and Liu, Y. The journals, Computers and Electronics in Agriculture and Sensors are among the most cited sources (Figs. 1.28 and 1.30 and Tables 1.28 and 1.30).

1.3   Bibliometric Analysis for the Topic "Agricultur* and Food 4.0"

In this topic 5 and 7, documents were found on the WoS and Scopus platforms, respectively. Considering the reduced number of documents for this topic, more attention was given to the links co-occurrence and items all and author keywords, taking into account the pertinence of this information for stakeholders who are involved in this topic.

Considering the information obtained from WoS, it is possible to identify 4 clusters for all keywords (Fig. 1.31) and 3 clusters for author keywords (Fig. 1.32). From the Scopus documents, 6 clusters were found for all keywords (Fig. 1.33) and 4 clusters were obtained for author keywords (Fig. 1.34). These clusters and the respective

Table 1.11 Top 20 most cited authors networked from WoS for citation links

Authors	Documents	Citations	Avg. pub. year
Klerkx, Laurens	3	230	2020
Jakku, Emma	1	128	2019
Labarthe, Pierre	1	128	2019
Rose, David Christian	2	116	2020
Chilvers, Jason	1	87	2018
Rovira-Mas, Francisco	1	81	2020
Saiz-Rubio, Veronica	1	81	2020
Alemany Diaz, Maria Del Mar Eva	1	77	2020
Hernandez, Jorge e	1	77	2020
Kacprzyk, Janusz	1	77	2020
Lezoche, Mario	1	77	2020
Panetto, Herve	1	77	2020
Rose, David	1	73	2020
Cecchini, Massimo	1	69	2019
Colantoni, Andrea	1	69	2019
Egidi, Gianluca	1	69	2019
Saporito, Maria Grazia	1	69	2019
Zambon, Ilaria	1	69	2019
Beltran, Victoria	1	65	2020
Fernan Martinez, Jose	1	65	2020

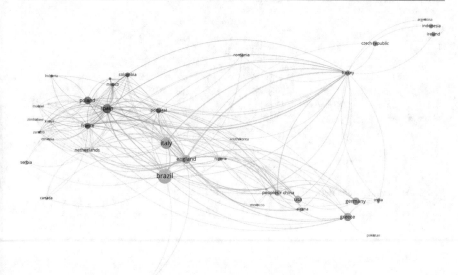

Fig. 1.12 Countries networked from WoS for citation links

Table 1.12 Top 20 most cited countries networked from WoS for citation links

Countries	Documents	Citations	Avg. pub. year
England	10	320	2020
Spain	12	266	2020
Netherlands	5	261	2020
France	5	205	2020
Australia	2	131	2020
Italy	18	101	2020
Poland	8	85	2020
Brazil	29	58	2020
Peoples r China	5	53	2021
Colombia	3	35	2020
Canada	2	34	2021
Greece	9	31	2021
South Africa	1	26	2021
Mexico	2	21	2021
Scotland	1	21	2020
Algeria	2	17	2021
South Korea	1	15	2018
USA	8	15	2020
Portugal	4	14	2021
Ethiopia	1	10	2020

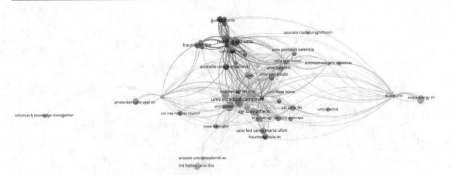

Fig. 1.13 Organisations networked from WoS for citation links

Table 1.13 Top 20 most cited organisations networked from WoS for citation links

Organisations	Documents	Citations	Avg. pub. year
Wageningen Univ	4	260	2020
Univ Politecn Valencia	2	158	2020
Csiro Land & Water	1	128	2019
Inra	1	128	2019
Univ Reading	3	103	2021
Univ East Anglia	1	87	2018
Syst Res Inst	1	77	2020
Univ Liverpool	1	77	2020
Univ Lorraine	1	77	2020
Minist Politiche Agr Alimentari & Forestali Mipaa	1	69	2019
Tuscia Univ	1	69	2019
Upm	2	67	2021
Nanjing Agr Univ	4	45	2021
Univ Lincoln	3	43	2021
Coll Social Sci & Int Studies	1	29	2021
Univ Ottawa	1	29	2020
Wageningen Univ & Res	1	29	2020
Chinese Acad Sci	1	26	2021
City Univ Hong Kong	1	26	2021
Univ Chinese Acad Sci	1	26	2021

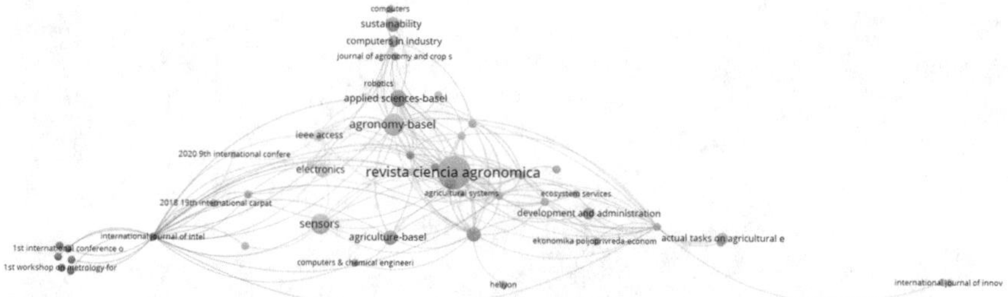

Fig. 1.14 Sources networked from WoS for citation links

Table 1.14 Top 20 most cited sources networked from WoS for citation links

Sources	Documents	Citations	Avg. pub. year
Njas-Wageningen Journal of Life Sciences	1	128	2019
Agronomy-Basel	7	91	2021
Frontiers in Sustainable Food Systems	1	87	2018
Processes	2	84	2019
Computers in Industry	2	77	2021
Global Food Security-Agriculture Policy Economics and Environment	1	73	2020

(continued)

Table 1.14   (continued)

Sources	Documents	Citations	Avg. pub. year
Computers and Electronics in Agriculture	3	68	2020
Applied Sciences-Basel	4	32	2021
Agricultural Systems	1	30	2020
Ecosystem Services	1	29	2020
Land Use Policy	1	29	2021
IEEE Transactions on Industrial Informatics	1	26	2021
IEEE Access	2	21	2021
Sensors	6	19	2021
2019 Global IoT Summit (Giots)	1	13	2019
IEEE-CAA Journal of Automatica Sinica	1	12	2021
Electronics	3	11	2021
Journal of Engineering	1	10	2020
Agriculture-Basel	3	9	2020
Computers & Chemical Engineering	1	9	2019

Fig. 1.15   Authors networked from Scopus for citation links

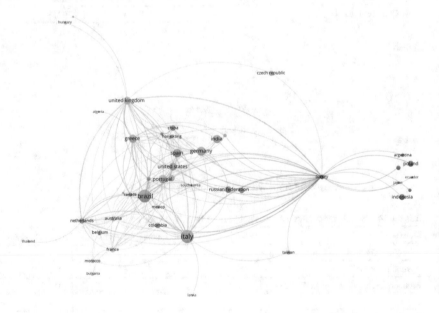

Fig. 1.16   Countries networked from Scopus for citation links

Fig. 1.17 Organisations networked from Scopus for citation links

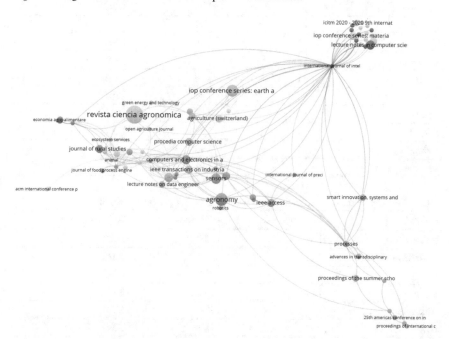

Fig. 1.18 Sources networked from Scopus for citation links

Table 1.15 Top 20 most cited authors networked from Scopus for citation links

Authors	Documents	Citations	Avg. pub. year
Klerkx l	5	331	2020
Jakku e	2	167	2020
Labarthe p	1	161	2019
Rose d.c	3	152	2020
Chilvers j	1	106	2018
Rovira-Más f	1	105	2020
Saiz-Rubio v	1	105	2020
Alemany díaz m.m.e	1	99	2020

(continued)

Table 1.15 (continued)

Authors	Documents	Citations	Avg. pub. year
Hernandez j.e	1	99	2020
Kacprzyk j	1	99	2020
Lezoche m	1	99	2020
Panetto h	1	99	2020
Cecchini m	1	96	2019
Colantoni a	1	96	2019
Egidi g	1	96	2019
Saporito m.g	1	96	2019
Zambon i	1	96	2019
Rose d	1	88	2020
Beltran v	1	83	2020
Martínez j.f	1	83	2020

Table 1.16 Top 20 most cited countries networked from Scopus for citation links

Countries	Documents	Citations	Avg. pub. year
Netherlands	6	334	2020
United kingdom	11	315	2020
France	6	260	2020
Spain	14	251	2020
Australia	6	174	2020
Italy	32	160	2020
Brazil	31	73	2020
Portugal	13	53	2020
Greece	12	52	2021
Colombia	6	47	2020
Malaysia	8	46	2020
China	6	42	2021
Germany	15	42	2019
Canada	3	39	2021
Hong Kong	2	31	2021
South Africa	3	31	2021
Poland	7	30	2020
Mexico	2	29	2021
United States	11	23	2020
Romania	6	22	2019

Table 1.17 Top 20 most cited organisations networked from Scopus for citation links

Organisations	Documents	Citations	Avg. pub. year
Knowledge, Technology and Innovation Group, Wageningen University, Netherlands	3	231	2020
Csiro Land and Water, Ecosciences Precinct Dutton Park, Queensland, Australia	1	161	2019
Inra, Umr Agir, Toulouse, France	1	161	2019
Science, Society and Sustainability (3 s) Research Group, School of Environmental Sciences, University of East Anglia, Norwich, United Kingdom	1	106	2018
Agricultural Robotics Laboratory (ARL), Universitat Politècnica de València, Spain	1	105	2020
University of Lorraine, Cran/Université de Lorraine, France	1	99	2020
Department of Agricultural and Forestry Sciences (Dafne), Tuscia University, Italy	1	96	2019
Ministero Delle Politiche Agricole Alimentari e Forestali (Mipaaf), Italy	1	96	2019
Sea Tuscia, Tuscia University, Italy	1	96	2019
Knowledge, Technology, and Innovation Group, Wageningen University, Netherlands	1	88	2020
School of Agriculture, Policy and Development, University of Reading, United Kingdom	1	88	2020
Departamento De Ingeniería Telemática Y Electrónica (DTE), Escuela Técnica Superior De Ingeniería Y Sistemas De Telecomunicación (Etsist), Universidad Politécnica De Madrid, Spain	1	83	2020
School of Sociological and Anthropological Studies, University Of Ottawa, Canada	1	31	2020
Wageningen Economic Research, Wageningen University and Research, Netherlands	1	31	2020
College of Engineering, Nanjing Agricultural University, China	1	30	2021
College of Science, University of Lincoln, United Kingdom	1	30	2021
Department of Computer Science, City University of Hong Kong, Hong Kong	1	30	2021
Department of Electrical and Electronic Engineering Science, Council for Scientific and Industrial Research, University of Johannesburg, South Africa	1	30	2021
School of Electronic, Electrical and Communication Engineering, University of Chinese Academy of Sciences, China	1	30	2021
Shanghai Advanced Research Institute, Chinese Academy of Sciences, China	1	30	2021

Table 1.18 Top 20 most cited sources networked from Scopus for citation links

Sources	Documents	Citations	Avg. pub. year
Njas-Wageningen Journal of Life Sciences	1	161	2019
Sgronomy	8	123	2021
Processes	2	117	2019
Frontiers in Sustainable Food Systems	1	106	2018
Computers in Industry	2	99	2021
Computers and Electronics in Agriculture	4	90	2021
Global Food Security	1	88	2020
Agricultural Systems	2	45	2021
Applied Sciences (Switzerland)	4	40	2021
Ecosystem Services	1	31	2020
IEEE Transactions on Industrial Informatics	3	31	2021
Land Use Policy	1	29	2021
IEEE Access	3	28	2021
Landtechnik	1	22	2016
Green Energy and Technology	1	17	2018
Sensors (Switzerland)	3	17	2020
Sociologia Ruralis	2	17	2021
Journal of Rural Studies	3	16	2021
Global IoT Summit, GIoTS 2019—Proceedings	1	12	2019
2018 13th International Conference on Digital Information Management, ICDIM 2018	1	11	2018

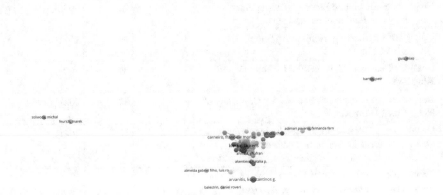

Fig. 1.19 Authors networked from WoS for bibliographic coupling links

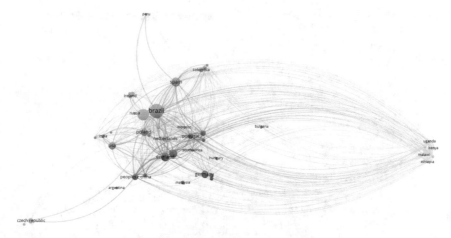

Fig. 1.20 Countries networked from WoS for bibliographic coupling links

Fig. 1.21 Organisations networked from WoS for bibliographic coupling links

distances between the several keywords highlight the relatedness among these items (number of documents in which the items occur together).

Keywords having more occurrences are the following (Tables 1.31, 1.32, 1.33 and 1.34): management; agri-food 4.0; decision-support-system; future; industry 4.0; IoT; sustainability; agriculture 4.0; agriculture supply chain; big data; agricultural robots; artificial intelligence; blockchain; digital technologies; biorefinery; and circular economy.

1.4 Bibliometric Analysis for the Topic "Agricultur* and Industry 4.0"

For the topic "agricultur* and industry 4.0", 157 and 342 documents were found, respectively, from WoS and Scopus. To avoid being too exhaustive, in this section, the information shown for the link bibliographic coupling was only considered using documents from the Scopus database (having more documents). In fact, the analysis carried out in the previous section reveals that there are some similarities in the

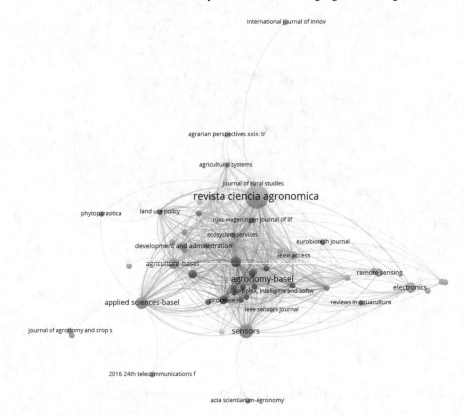

Fig. 1.22 Sources networked from WoS for bibliographic coupling links

Fig. 1.23 Authors networked from Scopus for bibliographic coupling links

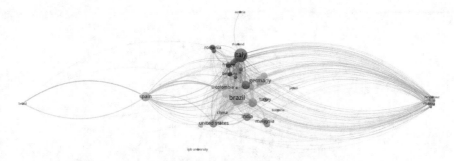

Fig. 1.24 Countries networked from Scopus for bibliographic coupling links

Fig. 1.25 Organisations networked from Scopus for bibliographic coupling links

findings for the different links and databases. In any case, these questions for this specific topic may be analysed further in future research.

The most cited authors are the following (Fig. 1.35 and Table 1.35): Alemany Díaz M.M.E.; Hernandez J.E.; Kacprzyk J.; Lezoche M.; and Panetto H. On the other hand, Italy, the United States, India, France and Taiwan are the most cited countries (Italy and India are also the countries with more documents) (Fig. 1.36 and Table 1.36). The organisations with more citations are the following (Fig. 1.37 and Table 1.37): University of Lorraine; Tuscia University; Ministero Delle Politiche Agricole Alimentari e Forestali (Italy); National Tsing Hua University (Taiwan); and Tecnologico de Monterrey, School of Engineering and Sciences (Mexico). Computers in Industry, IEEE Access, Processes, Computers and Electronics in Agriculture, and Robotics and Computer-Integrated Manufacturing are among the most cited sources (Fig. 1.38 and Table 1.38).

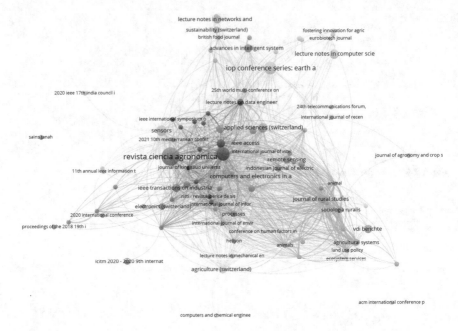

Fig. 1.26 Sources networked from Scopus for bibliographic coupling links

Table 1.19 Top 20 most cited authors networked from WoS for bibliographic coupling links

Authors	Documents	Citations	Avg. pub. year
Klerkx, Laurens	3	230	2019.667
Jakku, Emma	1	128	2019
Labarthe, Pierre	1	128	2019
Rose, David Christian	2	116	2019.5
Chilvers, Jason	1	87	2018
Rovira-Mas, Francisco	1	81	2020
Saiz-Rubio, Veronica	1	81	2020
Alemany Diaz, Maria Del Mar Eva	1	77	2020
Hernandez, Jorge e	1	77	2020
Kacprzyk, Janusz	1	77	2020
Lezoche, Mario	1	77	2020
Panetto, Herve	1	77	2020
Rose, David	1	73	2020
Cecchini, Massimo	1	69	2019
Colantoni, Andrea	1	69	2019
Egidi, Gianluca	1	69	2019
Saporito, Maria Grazia	1	69	2019

(continued)

Table 1.19 (continued)

Authors	Documents	Citations	Avg. pub. year
Zambon, Ilaria	1	69	2019
Beltran, Victoria	1	65	2020
Fernan Martinez, Jose	1	65	2020

Table 1.20 Top 20 most cited countries networked from WoS for bibliographic coupling links

Countries	Documents	Citations	Avg. pub. year
England	10	320	2020
Spain	12	266	2020
Netherlands	5	261	2020
France	5	205	2020
Australia	2	131	2020
Italy	18	101	2020
Poland	8	85	2020
Brazil	29	58	2020
Peoples r China	5	53	2021
Colombia	3	35	2020
Canada	2	34	2021
Greece	9	31	2021
South Africa	1	26	2021
Mexico	2	21	2021
Scotland	1	21	2020
Slgeria	2	17	2021
USA	8	15	2020
Portugal	4	14	2021
Malaysia	2	11	2019
Ethiopia	1	10	2020

Table 1.21 Top 20 most cited organisations networked from WoS for bibliographic coupling links

Organisations	Documents	Citations	Avg. pub. year
Wageningen Univ	4	260	2020
Univ Politecn Valencia	2	158	2020
Csiro Land & Water	1	128	2019
Inra	1	128	2019
Univ Reading	3	103	2021
Univ East Anglia	1	87	2018
Syst Res Inst	1	77	2020
Univ Liverpool	1	77	2020

(continued)

Table 1.21   (continued)

Organisations	Documents	Citations	Avg. pub. year
Univ Lorraine	1	77	2020
Minist Politiche Agr Alimentari & Forestali Mipaa	1	69	2019
Tuscia Univ	1	69	2019
Upm	2	67	2021
Nanjing Agr Univ	4	45	2021
Univ Lincoln	3	43	2021
Coll Social Sci & Int Studies	1	29	2021
Univ Ottawa	1	29	2020
Wageningen Univ & Res	1	29	2020
Chinese Acad Sci	1	26	2021
City Univ Hong Kong	1	26	2021
Univ Chinese Acad Sci	1	26	2021

Table 1.22   Top 20 most cited sources networked from WoS for bibliographic coupling links

Sources	Documents	Citations	Avg. pub. year
Njas-Wageningen Journal of Life Sciences	1	128	2019
Agronomy-Basel	7	91	2021
Frontiers in Sustainable Food Systems	1	87	2018
Processes	2	84	2019
Computers in Industry	2	77	2021
Global Food Security-Agriculture Policy Economics and Environment	1	73	2020
Computers and Electronics in Agriculture	3	68	2020
Applied Sciences-Basel	4	32	2021
Agricultural Systems	1	30	2020
Ecosystem Services	1	29	2020
Land Use Policy	1	29	2021
IEEE Transactions on Industrial Informatics	1	26	2021
IEEE Access	2	21	2021
Sensors	6	19	2021
2019 Global IoT Summit (Giots)	1	13	2019
IEEE-CAA Journal of Automatica Sinica	1	12	2021
Electronics	3	11	2021
Phytoparasitica	1	11	2019
Journal of Engineering	1	10	2020
Agriculture-Basel	3	9	2020

Table 1.23 Top 20 most cited authors networked from Scopus for bibliographic coupling links

Authors	Documents	Citations	Avg. pub. year
Klerkx l	5	331	2020
Jakku e	2	167	2020
Labarthe p	1	161	2019
Rose d.c	3	152	2020
Chilvers j	1	106	2018
Rovira-Más f	1	105	2020
Saiz-Rubio v	1	105	2020
Alemany Díaz m.m.e	1	99	2020
Hernandez j.e	1	99	2020
Kacprzyk j	1	99	2020
Lezoche m	1	99	2020
Panetto h	1	99	2020
Cecchini m	1	96	2019
Colantoni a	1	96	2019
Egidi g	1	96	2019
Saporito m.g	1	96	2019
Zambon i	1	96	2019
Rose d	1	88	2020
Beltran v	1	83	2020
Martínez j.f	1	83	2020

Table 1.24 Top 20 most cited countries networked from Scopus for bibliographic coupling links

Countries	Documents	Citations	Avg. pub. year
Netherlands	6	334	2020
United Kingdom	11	315	2020
France	6	260	2020
Spain	14	251	2020
Australia	6	174	2020
Italy	32	160	2020
Brazil	31	73	2020
Portugal	13	53	2020
Greece	12	52	2021
Colombia	6	47	2020
Malaysia	8	46	2020
China	6	42	2021
Germany	15	42	2019

(continued)

Table 1.24 (continued)

Countries	Documents	Citations	Avg. pub. year
Canada	3	39	2021
Hong Kong	2	31	2021
South Africa	3	31	2021
Poland	7	30	2020
Mexico	2	29	2021
United States	11	23	2020
Romania	6	22	2019

Table 1.25 Top 20 most cited organisations networked from Scopus for bibliographic coupling links

Organisations	Documents	Citations	Avg. pub. year
Knowledge, Technology and Innovation Group, Wageningen University, Netherlands	3	231	2020
CSIRO Land and Water, Ecosciences Precinct Dutton Park, Australia	1	161	2019
INRA, France	1	161	2019
Science, Society and Sustainability (3 S) Research Group, School of Environmental Sciences, University of East Anglia, United Kingdom	1	106	2018
Agricultural Robotics Laboratory (Arl), Universitat Politècnica De València, Spain	1	105	2020
University of Lorraine, France	1	99	2020
Department of Agricultural and Forestry Sciences (DAFNE), Tuscia University, Italy	1	96	2019
Ministero Delle Politiche Agricole Alimentari E Forestali (MIPAAF), Italy	1	96	2019
Sea Tuscia, Tuscia University, Italy	1	96	2019
Knowledge, Technology, and Innovation Group, Wageningen University, Netherlands	1	88	2020
School of Agriculture, Policy and Development, University of Reading, United Kingdom	1	88	2020
Departamento De Ingeniería Telemática Y Electrónica (Dte), Escuela Técnica Superior De Ingeniería Y Sistemas De Telecomunicación (Etsist), Universidad Politécnica De Madrid (UPM), Spain	1	83	2020
School of Sociological and Anthropological Studies, University of Ottawa, Canada	1	31	2020

(continued)

Table 1.25 (continued)

Organisations	Documents	Citations	Avg. pub. year
Wageningen Economic Research, Wageningen University and Research, Netherlands	1	31	2020
College of Engineering, Nanjing Agricultural University, China	1	30	2021
College of Science, University of Lincoln, United Kingdom	1	30	2021
Department of Computer Science, City University of Hong Kong, Hong Kong	1	30	2021
Department of Electrical and Electronic Engineering Science, Council for Scientific and Industrial Research, University of Johannesburg, South Africa	1	30	2021
School of Electronic, Electrical and Communication Engineering, University of Chinese Academy of Sciences, China	1	30	2021
Shanghai Advanced Research Institute, Chinese Academy of Sciences, China	1	30	2021

Table 1.26 Top 20 most cited sources networked from Scopus for bibliographic coupling links

Sources	Documents	Citations	Avg. pub. year
Njas-Wageningen Journal Of Life Sciences	1	161	2019
Agronomy	8	123	2021
Processes	2	117	2019
Frontiers in Sustainable Food Systems	1	106	2018
Computers in Industry	2	99	2021
Computers and Electronics in Agriculture	4	90	2021
Global Food Security	1	88	2020
Agricultural Systems	2	45	2021
Applied Sciences (Switzerland)	4	40	2021
Ecosystem Services	1	31	2020
IEEE Transactions on Industrial Informatics	3	31	2021
Land use Policy	1	29	2021
IEEE Access	3	28	2021
Green Energy and Technology	1	17	2018
Sensors (Switzerland)	3	17	2020
Sociologia Ruralis	2	17	2021
Journal of Rural Studies	3	16	2021
Global IoT Summit, Giots 2019-Proceedings	1	12	2019
2018 13th International Conference on Digital Information Management, ICDIM 2018	1	11	2018
Agriculture (Switzerland)	3	11	2020

Fig. 1.27 Authors networked from WoS for co-citation links

Table 1.27 Top 20 most cited authors networked from WoS for co-citation links

Authors	Citations
Klerkx, l	45
Rose, dc	45
Fao	32
Wolfert, s	25
Liu, y	19
Bronson, k	17
Carolan, m	17
Associacao Brasileira de Normas Tecnicas	16
Kamilaris, a	16
Zambon, i	15
De clercq, m	13
Eastwood, c	13
Shamshiri, rr	13
Zhai, zy	13
Lioutas, ed	12
European, Commission	11
Lezoche, m	11
Mancini, a	11
Bechar, a	10
Oecd	10

Fig. 1.28 Sources networked from WoS for co-citation links

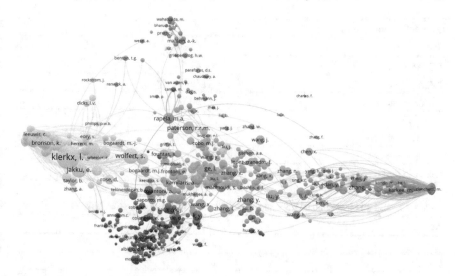

Fig. 1.29 Authors networked from Scopus for co-citation links

Fig. 1.30 Sources networked from Scopus for co-citation links

Table 1.28 Top 20 most cited sources networked from WoS for co-citation links

Sources	Citations
Comput Electron Agr	337
Sensors-Basel	158
Biosyst Eng	103
Agr Syst	101
Njas-Wagen J Life Sc	90
IEEE Access	82
Precis Agric	80
Sustainability-Basel	73
Remote Sens-Basel	63
Procedia Comput Sci	60
Int J Prod Res	55
Procedia Manuf	52
Comput Ind	49

(continued)

Table 1.28 (continued)

Sources	Citations
J Clean Prod	49
J Rural Stud	43
IFAC Papersonline	40
J Field Robot	39
Agr Water Manage	37
Agronomy-Basel	37
Land use Policy	36

Table 1.29 Top 20 most cited authors networked from Scopus for co-citation links

Authors	Citations
Klerkx, l	140
Rose, d.c	65
Jakku, e	63
Wolfert, s	61
Liu, y	40
Paterson, r.r.m	38
Chilvers, j	37
Verdouw, c	37
Bronson, k	36
Zhang, y	36
Rapela, m.a	32
Ge, l	31
Zhang, q	31
Carolan, m	30
Kamilaris, a	28
Labarthe, p	28
Shu, l	28
Wang, x	28
Eastwood, c	27
Li, y	27

Table 1.30 Top 20 most cited sources networked from Scopus for co-citation links

Sources	Citations
Computers and Electronics in Agriculture	193
Sensors	184
Comput. Electron. Agric	123
IEEE Access	106
Procedia Manufacturing	82
Comput. Electron. Agric	78
Sustainability	74
Agric. Syst	67
Njas-Wageningen j. Life sci	60
Biosystems Engineering	56
Procedia Computer Science	55
Biosyst. Eng	49
Precis. Agric	44
Agronomy	43
Precision Agriculture	43
Engenharia Agricola	41
Agricultural Systems	39
Land use Policy	39
Remote Sens	36
Remote Sensing	36

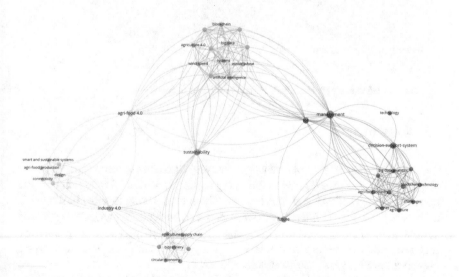

Fig. 1.31 All keywords networked from WoS for co-occurrence links

Fig. 1.32 Author keywords networked from WoS for co-occurrence links

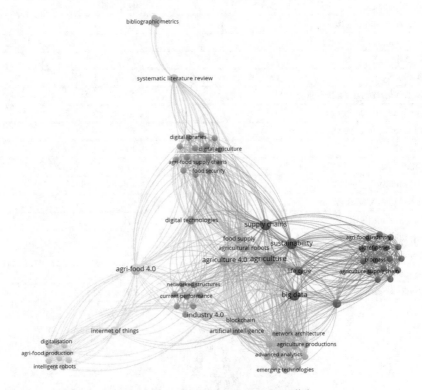

Fig. 1.33 All keywords networked from Scopus for co-occurrence links

1.5 Conclusions

The main objective of this research was to analyse bibliographic information from the WoS and Scopus databases for the following topics: agricultur* 4.0; agricultur* and food 4.0; and agricultur* and industry 4.0. The intention was to show bibliometric findings for the diverse interrelationships between agriculture and other sectors in the context of Era 4.0. To achieve these objectives, several documents were obtained from WoS and Scopus database that were analysed through bibliometric analysis, following the VOSviewer software procedures.

Among the topics considered, the results reveal that the subject which motivates researchers the most is "agricultur* and industry 4.0". In fact, for this topic, 157 and 342 studies were found, respectively, in the WoS and Scopus databases. The results

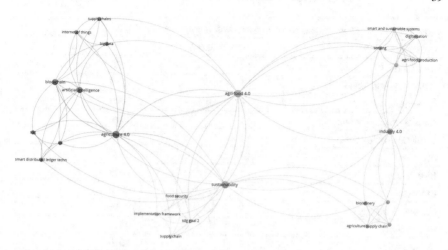

Fig. 1.34 Author keywords networked from Scopus for co-occurrence links

Table 1.31 Top 20 all keywords networked, with more occurrences, from WoS for co-occurrence links

All keywords	Occurrences	Avg. pub. year
Management	3	2020
Agri-food 4.0	2	2020
Decision-support-system	2	2021
Future	2	2020
Industry 4.0	2	2019
IoT	2	2021
Sustainability	2	2020
Agri-food industry	1	2021
Agri-food production	1	2019
Agriculture	1	2021
Agriculture 4.0	1	2020
Agriculture supply chain	1	2019
Artificial intelligence	1	2020
Artificial intelligence (ai)	1	2021
Big data	1	2020
Big data analytics	1	2021
Biorefinery	1	2019
Blockchain	1	2020
Blockchain technology	1	2021
Challenges	1	2021

Table 1.32 Author keywords networked from WoS for co-occurrence links

Author keywords	Occurrences	Avg. pub. year
Agri-food 4.0	2	2020
Agri-food production	1	2019
Agriculture 4.0	1	2020
Agriculture supply chain	1	2019
Artificial intelligence	1	2020
Big data	1	2020
Biorefinery	1	2019
Blockchain	1	2020
Circular economy	1	2019
Connectivity	1	2019
Digitalisation	1	2019
Industry 4.0	2	2019
Internet of things	1	2020
Life-cycle assessment	1	2019
Sensing	1	2019
Smart and sustainable systems	1	2019
Supply chains	1	2020
Sustainability	1	2019

Table 1.33 Top 20 all keywords networked, with more occurrences, from Scopus for co-occurrence links

All keywords	Occurrences	Avg. pub. year
Agriculture	4	2020
Agri-food 4.0	3	2020
Agriculture 4.0	3	2021
Big data	3	2020
Industry 4.0	3	2019
Supply chains	3	2020
Sustainability	3	2020
Agricultural robots	2	2021
Artificial intelligence	2	2021
Blockchain	2	2021
Digital technologies	2	2021
Food supply	2	2021
Internet of things	2	2020

(continued)

Table 1.33 (continued)

All keywords	Occurrences	Avg. pub. year
Life cycle	2	2020
Sustainable development	2	2020
Systematic literature review	2	2021
Advanced analytics	1	2021
Agri-food industry	1	2019
Agri-food production	1	2019
Agri-food supply chains	1	2021

Table 1.34 Top 20 author keywords networked, with more occurrences, from Scopus for co-occurrence links

Author keywords	Occurrences	Avg. pub. year
Agri-food 4.0	3	2020
Agriculture 4.0	3	2021
Sustainability	3	2020
Artificial intelligence	2	2021
Blockchain	2	2021
Industry 4.0	2	2019
Agri-food production	1	2019
Agriculture supply chain	1	2019
Big data	1	2020
Biorefinery	1	2019
Circular economy	1	2019
Connectivity	1	2019
Digitalisation	1	2019
Food security	1	2021
Implementation framework	1	2021
Internet of things	1	2020
IoT	1	2021
Life-cycle assessment	1	2019
Sdg	1	2021
Sdg goal 2	1	2021

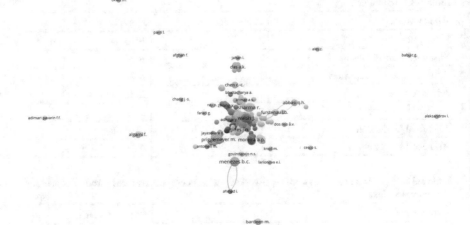

Fig. 1.35 Authors networked from Scopus for bibliographic coupling links

Table 1.35 Top 20 most cited authors networked from Scopus for bibliographic coupling links

Authors	Documents	Citations	Avg. pub. year
Alemany Díaz m.m.e	1	99	2020
Hernandez j.e	1	99	2020
Kacprzyk j	1	99	2020
Lezoche m	1	99	2020
Panetto h	1	99	2020
Cecchini m	1	96	2019
Colantoni a	1	96	2019
Egidi g	1	96	2019
Saporito m.g	1	96	2019
Zambon i	1	96	2019
Chuang a.c	1	95	2016
Govindarajan u.h	1	95	2016
Sun j.j	1	95	2016
Trappey a.j.c	1	95	2016
Trappey c.v	1	95	2016
Sharma r	4	77	2021
Miranda j	1	69	2019
Molina a	1	69	2019
Ponce p	1	69	2019
Wright p	1	69	2019

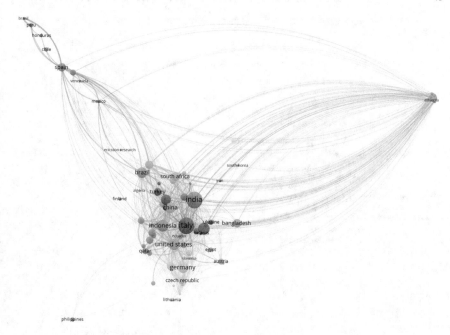

Fig. 1.36 Countries networked from Scopus for bibliographic coupling links

Table 1.36 Top 20 most cited countries networked from Scopus for bibliographic coupling links

Countries	Documents	Citations	Avg. pub. year
Italy	40	406	2020
United states	18	334	2020
India	40	262	2020
France	8	232	2020
Taiwan	8	136	2019
China	15	131	2020
Greece	6	107	2020
Australia	9	102	2020
Mexico	2	97	2020
Spain	10	83	2020
New Zealand	3	82	2019
Brazil	17	78	2020
United Kingdom	10	77	2020
Denmark	6	73	2020
Ireland	7	73	2019
South Africa	11	67	2020
Colombia	5	55	2020
Morocco	1	55	2020
Portugal	15	53	2020
Germany	18	43	2020

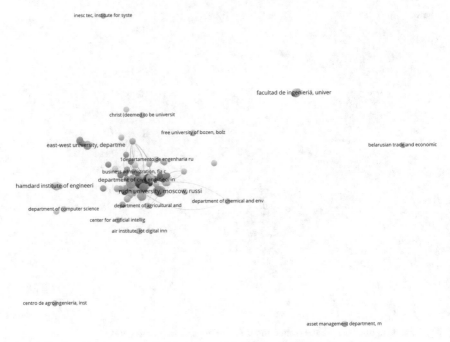

Fig. 1.37 Organisations networked from Scopus for bibliographic coupling links

also show that there is greater scope for further investigation in the topic "agricultur* and food 4.0" (5 and 7 documents were found on this topic in WoS and Scopus).

For the topic "agricultur* 4.0", the keywords with greater occurrence are the following: agriculture 4.0; precision agriculture; big data; internet of things; smart farming; agricultural robots; and industry 4.0. In the topic "agricultur* and food 4.0", keywords found with greater occurrence are the following: agri-food 4.0; decision-support-system; industry 4.0; IoT; sustainability; agriculture 4.0; agriculture supply chain; big data; agricultural robots; artificial intelligence; blockchain; digital technologies; biorefinery; and circular economy. In these topics, the keywords considered reveal signs of interrelationships between agriculture, industry and new technologies in order to deal with sustainability goals, namely in the last topic.

Considering the information for the topic "agricultur* 4.0" among the most cited authors are the following: Klerkx, Laurens; Jakku, Emma; Labarthe, Pierre; Rose, David Christian; and Chilvers, Jason. The Netherlands, United Kingdom, France, Spain and Australia are among the most cited countries. Among some of the most cited organisations, there are, for example: Wageningen Univ; Univ Politecn Valencia; Csiro Land and Water; Inra; and Univ Reading. Among the most cited sources, there are the following: Njas-Wageningen Journal of Life Sciences; Agronomy-Basel; Frontiers in Sustainable Food Systems; Processes; Computers in Industry; Computers and Electronics in Agriculture and Sensors. Some of these items also appear as the most cited for the topic "agricultur* and industry 4.0", nonetheless,

Table 1.37 Top 20 most cited organisations networked from Scopus for bibliographic coupling links

Organisations	Documents	Citations	Avg. pub. year
University of Lorraine, France	1	99	2020
Department of Agricultural and Forestry Sciences (Dafne), Tuscia University, Italy	1	96	2019
Ministero Delle Politiche Agricole Alimentari e Forestali (MIPAAF), Italy	1	96	2019
Sea Tuscia, Tuscia University, Italy	1	96	2019
Department of Industrial Engineering and Engineering Management, National Tsing Hua University, Taiwan	1	95	2016
Department of Management Science, National Chiao Tung University, Taiwan	1	95	2016
Tecnologico de Monterrey, School of Engineering and Sciences, Mexico	1	69	2019
University of California, United States	1	69	2019
Department of Electronics and Communication Engineering, Visvesvaraya National Institute of Technology Nagpur, India	1	67	2019
Laboratoire de Chimie Agro-Industrielle, Université de Toulouse, Inra, France	1	57	2019
Laboratoire de Génie Chimique, Université de Toulouse, France	1	57	2019
Cadi Ayyad University, Morocco	1	55	2020
Operations and Scm, Edhec Business School, France	1	55	2020
Operations Scm, Im Thapar School of Management, India	1	55	2020
Operations Scm, National Institute of Industrial Engineering (NITIE), India	1	55	2020
School of Business and Public Administration, California State University, United States	1	55	2020
Industrial Engineering and Automation (IEA), Faculty of Science and Technology, Free University of Bolzano, Italy	1	47	2021
Climate Change Cluster (C3), University of Technology Sydney, Australia	1	46	2020
Csiro Synthetic Biology Future Science Platform, Australia	1	46	2020
Department of Business Administration, Technological Educational Institute of Crete, Greece	1	44	2018

Fig. 1.38 Sources networked from Scopus for bibliographic coupling links

Table 1.38 Top 20 most cited sources networked from Scopus for bibliographic coupling links

Sources	Documents	Citations	Avg. pub. year
Computers in Industry	3	225	2019
IEEE Access	3	136	2018
Processes	3	105	2020
Computers and Electronics in Agriculture	3	99	2019
Robotics and Computer-Integrated Manufacturing	8	80	2021
IEEE Transactions on Industrial Informatics	4	77	2021
Industrial Robot	8	76	2020
Procedia Computer Science	11	70	2020
Sensors (Switzerland)	3	65	2020
International Journal of Logistics Research and Applications	1	55	2020
Resources, Conservation and Recycling	2	51	2019
Frontiers in Plant Science	1	46	2020
Lecture Notes on Data Engineering and Communications Technologies	3	44	2020
2016 27th Irish Signals and Systems Conference, ISSC 2016	1	43	2016
Journal of Imaging	1	39	2019
British Food Journal	1	32	2019
Procedia Manufacturing	4	32	2019
Competitiveness Review	1	31	2020
Journal of Cleaner Production	2	29	2021
Wireless Personal Communications	2	21	2020

in general, there are significant differences in the results for this topic relative to the topic "agricultur* 4.0".

In terms of practical implications, the recommendation is to promote further research concerning the topic "agricultur* and food 4.0", as there is still great potential that is yet to be explored here. For future studies, it could be interesting to consider these findings so as to carry out a systematic review covering these subjects.

Acknowledgements This work is funded by National Funds through the FCT—Foundation for Science and Technology, I.P., within the scope of the project Refª UIDB/00681/2020. Furthermore we would like to thank the CERNAS Research Centre and the Polytechnic Institute of Viseu for their support.

References

1. R.P. da Silva, A.F. dos Santos, B.R. de Oliveira, J.B. Costa Souza, D.T. de Oliveira, F.M. Carneiro, Potential of using statistical quality control in agriculture 4.0. Rev. Cienc. Agron. **51**, e20207745 (2020)
2. V.J. Pereira Domingues Martinho, Agri-food contexts in mediterranean regions: contributions to better resources management. Sustainability **13**, 6683 (2021)
3. M.K. Sott, L.B. Furstenau, L.M. Kipper, F.D. Giraldo, J.R. Lopez-Robles, M.J. Cobo, A. Zahid, Q.H. Abbasi, M.A. Imran, Precision techniques and agriculture 4.0 technologies to promote sustainability in the coffee sector: state of the art, challenges and future trends. IEEE Access **8**, 149854 (2020)
4. V.J.P.D. Martinho, Food and consumer attitude(s): an overview of the most relevant documents. Agricul. Basel **11**, 1183 (2021)
5. D.O. Aghimien, C.O. Aigbavboa, A.E. Oke, W.D. Thwala, Mapping out research focus for robotics and automation research in construction-related studies a bibliometric approach. J. Eng. Des. Technol. **18**, 1063 (2020)
6. M. Norouzi, M. Chafer, L.F. Cabeza, L. Jimenez, D. Boer, Circular economy in the building and construction sector: a scientific evolution analysis. J. Build. Eng. **44**, 102704 (2021)
7. G. Atzeni, G. Vignali, L. Tebaldi, E. Bottani, *A Bibliometric Analysis on Collaborative Robots in Logistics 4.0 Environments*, in *Proceedings of the 2nd International Conference on Industry 4.0 and Smart Manufacturing (Ism 2020)*, ed. by F. Longo, M. Affenzeller, A. Padovano, vol. 180 (Elsevier Science Bv, Amsterdam, 2021), pp. 686–695
8. G.P. Agnusdei, V. Elia, M.G. Gnoni, Is digital twin technology supporting safety management? a bibliometric and systematic review. Appl. Sci. Basel **11**, 2767 (2021)
9. A. Ahmi, H. Elbardan, R.H.R.M. Ali, *Bibliometric Analysis of Published Literature on Industry 4.0*, in *2019 International Conference on Electronics, Information, and Communication (Iceic)* (Ieee, New York, 2019), pp. 213–218
10. S. Echchakoui, N. Barka, Industry 4.0 and its impact in plastics industry: a literature review. J. Ind. Inf. Integr. **20**, 100172 (2020)
11. M.R. Nascimento, Knowledge ecosystems on Industry 4.0 in Brazil: a bibliometric analysis. A to Z **10**, (2021)
12. A. De las Heras, F. Relinque-Medina, F. Zamora-Polo, A. Luque-Sendra, Analysis of the evolution of the sharing economy towards sustainability. Trends and transformations of the concept. J. Clean Prod. **291**, 125227 (2021)
13. F. Garrigos-Simon, S. Sanz-Blas, Y. Narangajavana, D. Buzova, The nexus between big data and sustainability: an analysis of current trends and developments. Sustainability **13**, 6632 (2021)

14. H. Gholami, F. Abu, J.K.Y. Lee, S.S. Karganroudi, S. Sharif, Sustainable manufacturing 4.0-pathways and practices. Sustainability **13**, 13956 (2021)

15. M.A. Hoque, R. Rasiah, F. Furuoka, S. Kumar, Technology adoption in the apparel industry: insight from literature review and research directions. Res. J. Text. Appar. **25**, 292 (2021)

16. R. Miskiewicz, Internet of things in marketing: bibliometric analysis. Mark. Manag. Innov. 371 (2020)

17. D.A. Morelli, P.S.A. de Ignacio, Assessment of researches and case studies on cloud manufacturing: a bibliometric analysis. Int. J. Adv. Manuf. Technol. **117**, 691 (2021)

18. B. Pham-Duc, T. Tran, H.-T.-T. Le, N.-T. Nguyen, H.-T. Cao, T.-T. Nguyen, Research on industry 4.0 and on key related technologies in vietnam: a bibliometric analysis using scopus. Learn. Publ. **34**, 414 (2021)

19. A. Sierra-Henao, A. Munoz-Villamizar, E. Solano-Charris, J. Santos, *Sustainable Development Supported by Industry 4.0: A Bibliometric Analysis*, in *Service Oriented, Holonic and Multi-Agent Manufacturing Systems for Industry of the Future*, ed. by T. Borangiu, D. Trentesaux, P. Leitao, A.G. Boggino, V. Botti, vol. 853 (Springer International Publishing Ag, Cham, 2020), pp. 366–376

20. M. Lopes, R. Martins, Mapping the impacts of industry 4.0 on performance measurement systems. IEEE Latin Am. Trans. **19**, 1912 (2021)

21. Web of Science, *Web of Science Core Collection Database*. https://www.webofscience.com/wos/woscc/basic-search

22. Scopus, *Scopus Database*. https://www.scopus.com/search/form.uri?display=basic#basic

23. N. J. van Eck and L. Waltman, *Manual for VOSviewer Version 1.6.17*.

24. VOSviewer, *VOSviewer Software*. https://www.vosviewer.com//.

25. S. Türkeli, R. Kemp, B. Huang, R. Bleischwitz, W. McDowall, Circular economy scientific knowledge in the european union and china: a bibliometric, network and survey analysis (2006–2016). J. Clean. Prod. **197**, 1244 (2018)

26. V.J.P.D. Martinho, Output impacts of the single payment scheme in portugal: a regression with spatial effects. Outlook Agric **44**, 109 (2015)

27. V.J.P.D. Martinho, Insights from over 30 years of common agricultural policy in portugal. Outlook Agric **46**, 223 (2017)

Chapter 2
Systematic Review of Agriculture and Era 4.0: The Most Relevant Insights

2.1 Introduction

Agriculture 4.0 technologies appear to provide hope in dealing with the several challenges faced by farms nowadays. These tasks are related to the need for increasing efficiency, using resources in a sustainable way, dealing with global warming and reducing waste and residues [1]. In any case, these new approaches are not always easy to implement and the constraints are diverse [2]. These technologies, some of which are associated with the IoT (Internet of Things) and machine learning [3], among others, may contribute towards reducing food insecurity and improving sustainability [4].

The food markets are complex and often impacted upon by several particularities associated with demand and supply [5]. The way agri-food chains are organised and structured is decisive in the competitiveness of the agricultural sector and in achieving the food demand requirements worldwide. Digital technologies, from a perspective of Food 4.0, may bring relevant added value to the dynamics of food supply [6].

The new technologies of Era 4.0 provide innovative approaches to the interrelationships between the several dimensions of society and the economic sectors [7]. For example, the transformations brought about by Industry 4.0 in industrial contexts [8] and business contexts [9] and models [10] may contribute to a more sustainable food sector [11], from the perspective of a more circular economy [12].

Industry 4.0 relative to the autonomy of machines [13] and associated with digital transformation [14], as a crucial tool for business organisation and management [15], has attracted attention [16] from researchers and industry professionals [17] worldwide. This is currently an unavoidable approach [18] and a new tendency concept [19] that benefits from technological advances [20]. In these contexts, artificial intelligence has an important role to play in the competitiveness of companies [21] and the organisation of territories, from a perspective of smart factories [22], smart products, processes [23] and smart cities [24]. Industry 4.0 has promoted relevant structural changes in the manufacturing industry [25]. Typical questions for the different stake-holders are the relationships between Industry 4.0 and sustainability [26], production, operations management [27] and data management [28]. Environmental sustainability appears as a subject area in some studies in conjunction with the Industry 4.0 expression [29]. Industry 4.0 is often shown as having positive impacts on the state of the environment [30]. Nonetheless, the new waves of Industry 4.0 require investment and the respective budget, where structural policies may prove determinant [31].

In this way, the intention of this research is to highlight insights from the literature about the interrelationships between Era 4.0 and agriculture. In other words, the objective is to present contributions from the academic community covering the impacts of Era 4.0 on the agricultural sector and the influence of agriculture on wave 4.0. For this objective, a systematic review was carried out, following the PRISMA statement [32] complemented with bibliometric analysis (following a methodology known as MB2MBA2) [33]. For this systematic review, 7, 218 and 342 documents were obtained from the Scopus database [34] (has more documents than Web of Science), respectively, for the following topics: "agricultur* and food 4.0", "agricultur* 4.0" and "agricultur* and industry 4.0". For the bibliometric analysis, VOSviewer software was used [35, 36].

2.2   Bibliometric Analysis

In this section, terms that have more occurrences (for binary counting, occurrences represent the number of documents in which a term occurs) were first identified in the documents found for each topic, in order to highlight trends and relevance of expressions, and subsequently, the most relevant documents (most cited studies following [5]) were determined so as to be surveyed through systematic literature review.

2.2.1   Text Data, Co-occurrence Links and Terms as Items

Considering binary counting and one as the minimum number of occurrences of a term, the following figures exhibit the most relevant terms (with more occurrences) networked, respectively, for the topics "agricultur* and food 4.0"; "agricultur* 4.0"; and "agricultur* and industry 4.0". In the figures, the dimension of the circle associated with each term is relative to the number of occurrences, and the distance between items represents the relatedness.

Figure 2.1 and Table 2.1 show that for the topic "agricultur* and food 4.0" among the most relevant terms expressions such as the following appear: food, data, development, process, web, Africa, agri-food perspective, agricultural drone, agricultural waste and agriculture supply chain. These terms demonstrate that the new approaches from wave 4.0 have opened up new perspectives for the agri-food sector, namely in

Fig. 2.1 Terms networked from Scopus for co-occurrence links and the topic "agricultur* and food 4.0"

Table 2.1 Top 20 terms networked, with more occurrences, from Scopus for co-occurrence links and the topic "agricultur* and food 4.0"

Term	Occurrences	Avg. pub. year
Food	6	2020
Data	4	2020
Development	3	2020
Study	3	2020
Process	2	2020
Production	2	2020
Web	2	2021
Work	2	2020
Advantage	1	2019
Africa	1	2021
Agri-food perspective	1	2019
Agri-food production	1	2019
Agri-food sector	1	2019
Agri-food supply chain	1	2021
Agri-food supply chain stage	1	2021
Agricultural drone	1	2019
Agricultural waste	1	2019
Agricultural waste valorisation	1	2019
Agriculture supply chain	1	2020
Article	1	2019

terms of sustainability and dealing with problems of food insecurity in countries having greater difficulties.

For the topic "agricultur* 4.0", terms related to data and countries having more problems concerning food security appear again (Fig. 2.2 and Table 2.2). This shows the importance of the information for Agriculture 4.0 and the opportunities that this new Era may bring for the countries which have greater challenges in terms of agricultural sustainability. On the other hand, intelligence appears as the term with more occurrences for the topic "agricultur* and industry 4.0" (Fig. 2.3 and Table 2.3), namely artificial intelligence which is crucial nowadays for the competitiveness of companies.

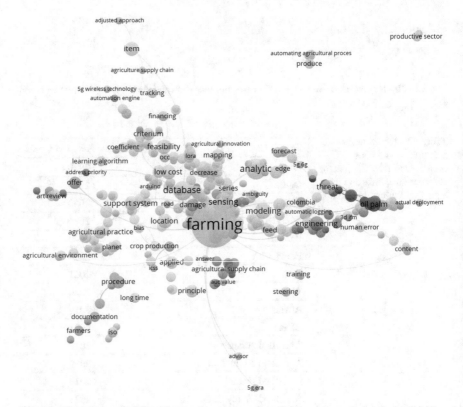

Fig. 2.2 Terms networked from Scopus for co-occurrence links and the topic "agricultur* 4.0"

2.2.2   The Most Cited Documents

To identify the most relevant documents (the most cited) for each one of the topics considered here ("agricultur* and food 4.0", "agricultur* 4.0" and "agricultur* and industry 4.0"), bibliographic coupling links were considered. For these links, the relatedness refers to the number of references the items share. The VOSviewer also allows for the identification of documents as items for the links citation (the relatedness is associated with the number of times the items cite each other), nonetheless the information obtained from the links' bibliographic coupling seems sufficient for the objectives set in this study. These documents, presented in Tables 2.4, 2.5 and 2.6, will be considered for systematic review in the next section. Some documents appear in more than one topic, showing that these subjects are interconnected.

Table 2.2 Top 20 terms networked, with more occurrences, from Scopus for co-occurrence links and the topic "agricultur* 4.0"

Terms	Occurrences	Avg. pub. year
Farming	41	2020
Analytic	8	2021
Database	7	2021
Sensing	6	2020
Modelling	5	2020
Artificial neural network	4	2021
Cattle	4	2020
Engineering	4	2020
Feasibility	4	2021
Item	4	2021
Location	4	2021
Low cost	4	2021
Monitoring system	4	2020
Oil palm	4	2020
Support system	4	2020
Threat	4	2020
Africa	3	2021
Agricultural practice	3	2021
Applied	3	2021
Behaviour	3	2020

Fig. 2.3 Terms networked from Scopus for co-occurrence links and the topic "agricultur* and industry 4.0"

Table 2.3 Top 20 terms networked, with more occurrences, from Scopus for co-occurrence links and the topic "agricultur* and industry 4.0"

Term	Occurrences	Avg. pub. year
Intelligence	34	2020
Generation	12	2020
Physical system	12	2020
Neural network	10	2021
Century	7	2020
Planning	7	2020
Range	7	2020
Shift	7	2019
Characteristic	6	2020
Entity	6	2020
Experimental result	6	2020
Machinery	6	2020
Age	5	2020
Class	5	2019
Controller	5	2019
Printing	5	2020
Regulation	5	2020
Visualisation	5	2019
Camera	4	2020
Capacity	4	2020

Table 2.4 Networked documents for bibliographic coupling links from Scopus for the topic "agricultur* and food 4.0"

Document	DOI	Citations
Eashwar [37]	https://doi.org/10.1088/1755-1315/775/1/012011	0
Oruma [4]	https://doi.org/10.1109/access.2021.3086453	0
Lezoche [38][a]	https://doi.org/10.1016/j.compind.2020.103187	99
Panetto [39]	https://doi.org/10.1016/j.compind.2020.103188	10
Belaud [40][b]	https://doi.org/10.1016/j.compind.2019.06.006	57
Miranda [41][b]	https://doi.org/10.1016/j.compind.2019.02.002	69

[a] Appears in the three topics
[b] Appear in the topics "agricultur* and food 4.0" and "agricultur* and industry 4.0"

Table 2.5 Top 20 most cited networked documents for bibliographic coupling links from Scopus for the topic "agricultur* 4.0"

Document	DOI	Citations
Klerkx [42]	https://doi.org/10.1016/j.njas.2019.100315	161
Rose [43]	https://doi.org/10.3389/fsufs.2018.00087	106
Saiz-Rubio [44]	https://doi.org/10.3390/agronomy10020207	105
Lezoche [45][a]	https://doi.org/10.1016/j.compind.2020.103187	99
Zambon [46][b]	https://doi.org/10.3390/pr7010036	96
Klerkx [47]	https://doi.org/10.1016/j.gfs.2019.100347	88
Zhai [1]	https://doi.org/10.1016/j.compag.2020.105256	83
Klerkx [47]	https://doi.org/10.1016/j.agsy.2020.102901	39
Lajoie-O'malley [48]	https://doi.org/10.1016/j.ecoser.2020.101183	31
Liu [49][b]	https://doi.org/10.1109/tii.2020.3003910	30
Rose [50]	https://doi.org/10.1016/j.landusepol.2020.104933	29
Sott [3][b]	https://doi.org/10.1109/access.2020.3016325	28
Symeonaki [51]	https://doi.org/10.3390/app10030813	24
Huh [52]	https://doi.org/10.3390/pr6090168	21
Barrett [53]	https://doi.org/10.1111/soru.12324	17
Yahya [54]	https://doi.org/10.1007/978-981-10-7578-0_5	17
Velásquez [55]	https://doi.org/10.3390/app10020697	15
Rijswijk [56]	https://doi.org/10.1016/j.jrurstud.2021.05.003	12
Borrero [57]	https://doi.org/10.3390/s20072078	12
Monteleone [58]	https://doi.org/10.1109/giots.2019.8766384	12

[a] Appears in the three topics
[b] Appear in the topics "agricultur* 4.0" and "agricultur* and industry 4.0"

2.3 Systematic Review

Tables 2.7, 2.8 and 2.9 highlight the main insights from the documents identified as the most cited for the topics "agricultur* and food 4.0", "agricultur* 4.0" and "agricultur* and industry 4.0". Some documents were removed from Tables 2.8 and 2.9 as they were duplicated, as shown in the previous section.

These tables show that the digital transformation and associated technologies, such as the Internet of Things, blockchain, robotics, artificial intelligence and big data analytics, are the basis of Era 4.0 (Agriculture 4.0, Food 4.0 and Industry 4.0). In this wave, several stakeholders are concerned with sustainability and believe that the associated innovations are able to promote more sustainable development in the sectors involved. Indeed, the digital transformation and smart approaches may provide an increase in the food supply without compromising the environment. Nonetheless, there are authors who discuss the negative impacts of Era 4.0, namely in terms of social sustainability, information privacy, costs associated and skills needed.

Table 2.6 Top 20 most cited networked documents for bibliographic coupling links from Scopus for the topic "agricultur* and industry 4.0"

Document	DOI	Citations
Lezoche [38][a]	https://doi.org/10.1016/j.compind.2020.103187	99
Zambon [45][c]	https://doi.org/10.3390/pr7010036	96
Trappey [59]	https://doi.org/10.1109/access.2016.2619360	95
Miranda [41][b]	https://doi.org/10.1016/j.compind.2019.02.002	69
Nawandar [60]	https://doi.org/10.1016/j.compag.2019.05.027	67
Belaud [40][b]	https://doi.org/10.1016/j.compind.2019.06.006	57
Sharma [61]	https://doi.org/10.1080/13675567.2020.1830049	55
Gualtieri [62]	https://doi.org/10.1016/j.rcim.2020.101998	47
Fabris [63]	https://doi.org/10.3389/fpls.2020.00279	46
Vassakis [64]	https://doi.org/10.1007/978-3-319-67925-9_1	44
Nukala [65]	https://doi.org/10.1109/issc.2016.7528456	43
Malik [66]	https://doi.org/10.1108/ir-11-2018-0231	42
Arachchige [67]	https://doi.org/10.1109/tii.2020.2974555	39
Mavridou [68]	https://doi.org/10.3390/jimaging5120089	39
Khan [69]	https://doi.org/10.3390/s20102990	37
Trivelli [70]	https://doi.org/10.1108/bfj-11-2018-0747	32
Dutta [71]	https://doi.org/10.1108/cr-03-2019-0031	31
Liu [49][c]	https://doi.org/10.1109/tii.2020.3003910	30
Sott [3][c]	https://doi.org/10.1109/access.2020.3016325	28
Martín-Gómez [72]	https://doi.org/10.1016/j.resconrec.2018.10.035	28

[a] Appears in the three topics
[b] Appear in the topics "agricultur* and food 4.0" and "agricultur* and industry 4.0"
[c] Appear in the topics "agricultur* 4.0" and "agricultur* and industry 4.0"

In any case, the literature reveals the relationships between Agriculture 4.0, Food 4.0 and Industry 4.0 because of the food supply chains, where the three sectors are involved and interconnected.

2.4   Conclusions

With this research, the intention was to show the main insights from the literature concerning the several interrelationships between agriculture and Era 4.0. To achieve this objective, a systematic review was carried out following the PRISMA statement complemented with the MB2MBA2 methodology. For the systematic review 7, 218 and 342 documents were found in the Scopus database, respectively, for the topics "agricultur* and food 4.0", "agricultur* 4.0" and "agricultur* and industry 4.0".

Table 2.7 Main insights from the documents obtained on Scopus for the topic "agricultur* and food 4.0", considering bibliographic coupling links

Document	Objectives/methodologies	Main insights
Eashwar [37]	Combine farming production with food distribution, considering the approaches from Industry 4.0	Internet of Things, Smart DLT and Big Data Analytics may be relevant technologies
Oruma [4]	Carry out a systematic literature review about Nigeria's agriculture and Agriculture 4.0	Agriculture 4.0 may contribute to achieve SDG goals
Lezoche [38]	Review about the new technologies and new supply chain methods in the agri-food 4.0 context	Smart agriculture and technologies associated with Internet of Things Blockchain, Big data and Artificial Intelligence may support improvements in the agri-food chains sustainability
Panetto [39]	Introduce the special issue about agri-food 4.0 and agricultural digitalisation	Agri-food chains involve complexes dimensions with several challenges
Belaud [40]	Use big data to valorise agricultural waste	It is possible to consider jointly information Industry 4.0, sustainability and agri-food chains
Miranda [41]	Consider the "sensing, smart and sustainable (S^3)" approach to achieve new technologies for the agri-food sector	The new product development process should integrate technologies, materials, processes and practices

The bibliometric analysis with text data, co-occurrence links and terms as items highlight that for the topic "agricultur* and food 4.0" new technologies bring new horizons to food supply chains, specifically in improving the sustainability of the activities involved and mitigating problems of food insecurity. On the other hand, the data appear as something relevant for the Agriculture 4.0 implementation (topic "agricultur* 4.0"). For the subjects related to "agricultur* and industry 4.0", smart approaches appear as the most relevant.

The systematic review, for these topics, shows that the digital transformation and associated technologies have been crucial in the transition from wave 3.0 to Era 4.0. The relationships of the sectors and activities involved (agriculture, food and industry) with sustainability are some of the main concerns of researchers, specifically over food supply chains. A relevant part of the scientific community highlights the positive contributions from Era 4.0 to sustainable development, but another part provides solid arguments concerning the negative impacts of this wave on social dimensions.

Table 2.8 Main insights from the top 20 most cited documents obtained on Scopus for the topic "agricultur* 4.0", considering bibliographic coupling links

Document	Objectives/methodologies	Main insights
Klerkxss [42]	Review about Agriculture 4.0	Several clusters on digital agriculture literature were identified and new topics on social sciences were suggested
Rose [43]	Discuss about responsible innovation in agriculture	It is important to test contexts in practice to assess their impacts
Saiz-Rubio [44]	Review about crop data management	Adjusted data management may support optimised decisions
Zambon [45]	Review agriculture vs. industry in Era 4.0	Policies may bring relevant contributions
Klerkx [46]	Analyse the dimensions of the Agriculture 4.0	It is important to anticipate impacts of Agriculture 4.0
Zhai [1]	Employ decision support systems in Agriculture 4.0	Several challenges were identified
Klerkx [47]	Review about innovation and policy sciences	Mission-oriented agricultural innovation systems solutions may bring relevant contributions
Lajoie-O'malley [48]	Analyse impacts of digital technologies on food systems	It could be important to investigate the environmental impacts of digital agriculture
Liu [49]	Review about the industrial agriculture	IoT, robotics, artificial intelligence, big data analytics and blockchain are emerging technologies in the context of Agriculture 4.0
Rose [50]	Discuss the several dimensions of Agriculture 4.0	The social sustainability has been neglected
Sott [3]	Systematic literature review about Agriculture 4.0 in the coffee sector	IoT, Machine Learning and geostatistics are important technologies in the coffee sector
symeonaki [51]	Integrate the precision agriculture into IoT	Innovative solutions were proposed
Huh [52]	Estimate the greenhouse gas emissions during the pig-manure composting	Agriculture 4.0 may support more sustainable approaches
Barrett [53]	Analyse technologies related to Agriculture 4.0	The new technologies have positive and negative impacts
Yahya [54]	Implement Agriculture 4.0 in a sustainable way	Agriculture 4.0 implementation may be challenging in some countries
Velásquez [55]	Detect Coffee Leaf Rust with new technologies	Remote sensing, wireless sensor networks and Deep Learning may be interesting solutions

(continued)

Table 2.8   (continued)

Document	Objectives/methodologies	Main insights
Rijswijk [56]	Understand impacts of the digital technologies on agriculture	Digital transformation has positive and negative impacts
Borrero [57]	Consider Internet of Things, wireless sensor networks and long-range wide-area network solutions to collect data	These solutions may improve the irrigation efficiency on farms
Monteleone [58]	Present a model to analyse variables that affect the adoption of practices in the framework of Agriculture 4.0	IoT and Industry 4.0 solutions may play a relevant role to optimise operations and resources

Table 2.9   Main insights from the top 20 most cited documents obtained on Scopus for the topic "agricultur* and industry 4.0", considering bibliographic coupling links

Document	Objectives/methodologies	Main insights
Trappey [59]	Review about Cyber-physical systems	The main trends of Industry 4.0 were highlighted
Nawandar [60]	Develop an IoT approach for smart irrigation	The proposed innovation is low cost, intelligent and portable
Sharma [61]	Identify agricultural supply chain risks created with the pandemic context	Industry 4.0 approaches may be useful to support stakeholders identifying solutions
Gualtieri [62]	Assess the design of ergonomic and safe industrial collaborative robotics	The main focus of the research is the safety. Nonetheless, in the recent years, the interest by the ergonomics increased
Fabris [63]	Review innovations in the domains of algal biotechnology	New technologies may promote the algae industrialisation with solutions for the society's requirements
Vassakis [64]	Discuss big data analytics	Big data is a determinant for business innovation and competitiveness
Nukala [65]	Use of IoT for food supply chain	The IoT technologies may be useful in food production, transportation, distribution and retail
Malik [66]	Present an approach for tasks sharing by human and robot in assembly work	The methodology separates the tasks with higher complexity from the low-complexity ones
Arachchige [67]	Introduce the PriModChain framework to improve the trust in the Industrial Internet of Things	This framework is compatible with the five pillars of a trustworthy system

(continued)

Table 2.9 (continued)

Document	Objectives/methodologies	Main insights
Mavridou [68]	Review documents about machine vision in agriculture	Agricultural robots for the harvest may be a solution for many contexts
Khan [69]	Help stakeholders of food industry to use the new technologies (IoT, blockchain and deep learning)	Consumers and producers may take advantage from a large amount of information
Trivelli [70]	Understand precision agriculture and Industry 4.0	The two approaches are directly interrelated
Dutta [71]	Study areas to promote Industry 4.0 in India	Manufactures are interested to carry out changes based on metric, but identify challenges
Martín-Gómez [72]	Propose a framework to promote a smart, integrated and sustainable social metabolism	This framework integrates the sustainable supply chains, considering the circular economy approaches and the pillars of sustainability

In terms of practical implications, Era 4.0 opens up several opportunities and challenges for several stakeholders in the agricultural sector, but this needs to be anticipated and properly assessed so as to guarantee an effective and adjusted implementation of these new technologies. The scientific community may provide a decisive contribution, but the involvement of farmers, food companies and rural entrepreneurs will be essential.

For policy recommendation, the suggestion would be that international organisations and governments design policies that support the adoption of these new technologies in a sustainable way, including for social dimensions. Smart approaches are important for the modernisation of the agricultural sector, but this involves costs, transformations and adaptations that call for institutional intervention in a farming sector with specific characteristics.

It could be interesting to consider the assessment of the international context in future studies, specifically in the European Union, through analysis of statistical information, which allows to identify strengths, weaknesses, opportunities and threats. These findings will support several stakeholders for more adjusted actions and decisions.

Acknowledgements This work is funded by National Funds through the FCT—Foundation for Science and Technology, I.P., within the scope of the project Ref^a UIDB/00681/2020. Furthermore we would like to thank the CERNAS Research Centre and the Polytechnic Institute of Viseu for their support.

References

1. Z. Zhai, J. Fernan Martinez, V. Beltran, N. Lucas Martinez, Decision support systems for agriculture 4.0: survey and challenges. Comput. Electron. Agric. **170**, 105256 (2020)
2. F. da Silveira, F.H. Lermen, F.G. Amaral, An overview of Agriculture 4.0 development: systematic review of descriptions, technologies, barriers, advantages, and disadvantages. Comput. Electron. Agric. **189**, 106405 (2021)
3. M.K. Sott, L.B. Furstenau, L.M. Kipper, F.D. Giraldo, J.R. Lopez-Robles, M.J. Cobo, A. Zahid, Q.H. Abbasi, M.A. Imran, Precision techniques and Agriculture 4.0 technologies to promote sustainability in the coffee sector: state of the art, challenges and future trends. IEEE Access **8**, 149854 (2020)
4. S.O. Oruma, S. Misra, L. Fernandez-Sanz, Agriculture 4.0: an implementation framework for food security attainment in nigeria's post-Covid-19 Era. IEEE Access **9**, 83592 (2021)
5. V.J.P.D. Martinho, Food and consumer attitude(s): an overview of the most relevant documents. Agricul. Basel **11**, 1183 (2021)
6. V. Dadi, S.R. Nikla, R.S. Moe, T. Agarwal, S. Arora, Agri-food 4.0 and innovations: revamping the supply chain operations. Prod. Eng. Arch. **27**, 75 (2021)
7. A.C.D. Aneze Ferreira, S.L. Meirelles Campos Titotto, A.C. Santos Akkari, *Urban Agriculture 5.0: An Exploratory Study*, in *2021 14th Ieee International Conference on Industry Applications (induscon)*, ed. by M.D.G. Tsuzuki, M.A.D. Pessoa (IEEE, New York, 2021), pp. 1441–1446
8. P. Morella, M.P. Lamban, J. Royo, J.C. Sanchez, Study and analysis of the implementation of 4.0 technologies in the agri-food supply chain: a state of the art. Agronomy Basel **11**, 2526 (2021)
9. J. Salvadorinho, L. Teixeira, Stories told by publications about the relationship between industry 4.0 and lean: systematic literature review and future research agenda. Publications **9**, 29 (2021)
10. J.E. Teixeira, A.T. Tavares-Lehmann, The confluence of i.4.0 technologies and new business models: a systematic literature review. Int. J. Innov. **9**, 664 (2021)
11. Y. Kayikci, N. Subramanian, M. Dora, M.S. Bhatia, Food supply chain in the era of industry 4.0: blockchain technology implementation opportunities and impediments from the perspective of people, process, performance, and technology. Prod. Plan. Control (n.d.)
12. K. Nayal, R. Raut, A.B.L.S. de Jabbour, B.E. Narkhede, V.V. Gedam, Integrated technologies toward sustainable agriculture supply chains: missing links. J. Enterp. Inf. Manag. (n.d.)
13. M.F. Manesh, M.M. Pellegrini, G. Marzi, M. Dabic, Knowledge management in the fourth industrial revolution: mapping the literature and scoping future avenues. IEEE Trans. Eng. Manag. **68**, 289 (2021)
14. M. Ghobakhloo, M. Fathi, M. Iranmanesh, P. Maroufkhani, M.E. Morales, Industry 4.0 ten years on: a bibliometric and systematic review of concepts, sustainability value drivers, and success determinants. J. Clean Prod. **302**, 127052 (2021)
15. X. Zhu, S. Ge, N. Wang, Digital transformation: a systematic literature review. Comput. Ind. Eng. **162**, 107774 (2021)
16. M. Habibi Rad, M. Mojtahedi, M.J. Ostwald, Industry 4.0, disaster risk management and infrastructure resilience: a systematic review and bibliometric analysis. Buildings Basel **11**, 411 (2021)
17. S. Tiwari, Supply chain integration and industry 4.0: a systematic literature review. Benchmarking **28**, 990 (2021)
18. P. Dhamija, Economic development and south africa: 25 years analysis (1994 to 2019). South Afr. J. Econ. **88**, 298 (2020)
19. J.A. Mesa, C. Fuquene-Retamoso, A. Maury-Ramirez, Life cycle assessment on construction and demolition waste: a systematic literature review. Sustainability **13**, 7676 (2021)
20. I. Taboada, H. Shee, Understanding 5G technology for future supply chain management. Int. J. Logist. Res. Appl. **24**, 392 (2021)
21. G. Buchgeher, D. Gabauer, J. Martinez-Gil, L. Ehrlinger, Knowledge graphs in manufacturing and production: a systematic literature review. IEEE Access **9**, 55537 (2021)

22. F. Strozzi, C. Colicchia, A. Creazza, C. Noe, literature review on the "smart factory" concept using bibliometric tools. Int. J. Prod. Res. **55**, 6572 (2017)

23. N.S. Zabidin, S. Belayutham, C.K.I.C. Ibrahim, A bibliometric and scientometric mapping of industry 4.0 in construction. J. Inf. Technol. Constr. **25**, (2020)

24. P. Tiwari, P.V. Ilavarasan, S. Punia, Content analysis of literature on big data in smart cities. Benchmarking **28**, 1837 (2021)

25. J. Lohmer, R. Lasch, Production planning and scheduling in multi-factory production networks: a systematic literature review. Int. J. Prod. Res. **59**, 2028 (2021)

26. K. Ejsmont, B. Gladysz, A. Kluczek, Impact of industry 4.0 on sustainability-bibliometric literature review. Sustainability **12**, 5650 (2020)

27. A. Felsberger, G. Reiner, Sustainable industry 4.0 in production and operations management: a systematic literature review. Sustainability **12**, 7982 (2020)

28. M.E. Hernandez Korner, M.P. Lamban, J.A. Albajez, J. Santolaria, L.C. del Ng Corrales, J. Royo, Systematic literature review: integration of additive manufacturing and industry 4.0, Metals **10**, 1061 (2020)

29. A. Corallo, M.E. Latino, M. Menegoli, P. Pontrandolfo, A systematic literature review to explore traceability and lifecycle relationship. Int. J. Prod. Res. **58**, 4789 (2020)

30. A. Farrukh, S. Mathrani, N. Taskin, Investigating the theoretical constructs of a green lean six sigma approach towards environmental sustainability: a systematic literature review and future directions. Sustainability **12**, 8247 (2020)

31. J. Wyrwa, A review of the european union financial instruments suppoting the innovative activity of enterprises in the context of industry 4.0 in the years 2021–2027, Entrep. Sustain. Iss. **8**, 1146 (2020)

32. A. Liberati, D.G. Altman, J. Tetzlaff, C. Mulrow, P.C. Gøtzsche, J.P.A. Ioannidis, M. Clarke, P.J. Devereaux, J. Kleijnen, D. Moher, The PRISMA statement for reporting systematic reviews and meta-analyses of studies that evaluate health care interventions: explanation and elaboration. PLOS Medicine **6**, e1000100 (2009)

33. V.J.P.D. Martinho, Agri-food contexts in mediterranean regions: contributions to better resources management. Sustainability **13**, 12 (2021)

34. Scopus, *Scopus Database*. https://www.scopus.com/search/form.uri?display=basic#basic

35. N.J. van Eck, L. Waltman, *Manual for VOSviewer Version 1.6.17*

36. VOSviewer, *VOSviewer Software*. https://www.vosviewer.com//

37. S. Eashwar, P. Chawla, Evolution of agritech business 4.0—architecture and future research directions. IOP Conf. Ser. Earth Environ. Sci. **775**, 012011 (2021)

38. M. Lezoche, J.E. Hernandez, M.E. del Alemany Díaz, H. Panetto, J. Kacprzyk, Agri-food 4.0: a survey of the supply chains and technologies for the future agriculture. Comput. Ind. **117**, 103187 (2020)

39. H. Panetto, M. Lezoche, J.E. Hernandez Hormazabal, M.E. del Alemany Diaz, J. Kacprzyk, Special issue on agri-food 4.0 and digitalization in agriculture supply chains-new directions, challenges and applications. Comput. Ind. **116**, 103188 (2020)

40. J.-P. Belaud, N. Prioux, C. Vialle, C. Sablayrolles, Big data for agri-food 4.0: application to sustainability management for by-products supply chain. Comput. Ind. **111**, 41 (2019)

41. J. Miranda, P. Ponce, A. Molina, P. Wright, Sensing, smart and sustainable technologies for agri-food 4.0. Comput. Ind. **108**, 21 (2019)

42. L. Klerkx, E. Jakku, P. Labarthe, A review of social science on digital agriculture, smart farming and agriculture 4.0: new contributions and a future research agenda. NJAS Wageningen J. Life Sci. **90–91**, 100315 (2019)

43. D.C. Rose, J. Chilvers, Agriculture 4.0: broadening responsible innovation in an era of smart farming. Front. Sustain. Food Syst. **2**, (2018)

44. V. Saiz-Rubio, F. Rovira-Más, From smart farming towards Agriculture 5.0: a review on crop data management. Agronomy **10**, 2 (2020)

45. I. Zambon, M. Cecchini, G. Egidi, M.G. Saporito, A. Colantoni, Revolution 4.0: industry versus agriculture in a future development for SMEs. Processes **7**, 1 (2019)

46. L. Klerkx, D. Rose, Dealing with the game-changing technologies of Agriculture 4.0: how do we manage diversity and responsibility in food system transition pathways?. Global Food Secur. **24**, 100347 (2020)
47. L. Klerkx, S. Begemann, Supporting food systems transformation: the what, why, who, where and how of mission-oriented agricultural innovation systems. Agricul. Syst. **184**, 102901 (2020)
48. A. Lajoie-O'Malley, K. Bronson, S. van der Burg, L. Klerkx, The future(s) of digital agriculture and sustainable food systems: an analysis of high-level policy documents. Ecosyst. Serv. **45**, 101183 (2020)
49. Y. Liu, X. Ma, L. Shu, G.P. Hancke, A.M. Abu-Mahfouz, From industry 4.0 to Agriculture 4.0: current status, enabling technologies, and research challenges. IEEE Trans. Ind. Infor. **17**, 4322 (2021)
50. D.C. Rose, R. Wheeler, M. Winter, M. Lobley, C.-A. Chivers, Agriculture 4.0: making it work for people, production, and the planet. Land Use Policy **100**, 104933 (2021)
51. E. Symeonaki, K. Arvanitis, D. Piromalis, A context-aware middleware cloud approach for integrating precision farming facilities into the IoT toward Agriculture 4.0. Appl. Sci. **10**, 3 (2020)
52. J.-H. Huh, K.-Y. Kim, Time-based trend of carbon emissions in the composting process of swine manure in the context of agriculture 4.0. Processes **6**, 9 (2018)
53. H. Barrett, D.C. Rose, Perceptions of the fourth agricultural revolution: what's in, what's out, and what consequences are anticipated?. Sociol. Ruralis n/a, (n.d.)
54. N. Yahya, Agricultural 4.0: Its Implementation Toward Future Sustainability, in *Green Urea : For Future Sustainability*. ed. by N. Yahya (Springer, Singapore, 2018), pp. 125–145
55. D. Velásquez, A. Sánchez, S. Sarmiento, M. Toro, M. Maiza, B. Sierra, A method for detecting coffee leaf rust through wireless sensor networks, remote sensing, and deep learning: case study of the caturra variety in colombia. Appl. Sci. **10**, 2 (2020)
56. K. Rijswijk, L. Klerkx, M. Bacco, F. Bartolini, E. Bulten, L. Debruyne, J. Dessein, I. Scotti, G. Brunori, Digital transformation of agriculture and rural areas: a socio-cyber-physical system framework to support responsibilisation. J. Rural. Stud. **85**, 79 (2021)
57. J.D. Borrero, A. Zabalo, An autonomous wireless device for real-time monitoring of water needs. Sensors **20**, 7 (2020)
58. S. Monteleone, E.A. de Moraes, R.F. Maia, *Analysis of the Variables That Affect the Intention to Adopt Precision Agriculture for Smart Water Management in Agriculture 4.0 Context*, in *2019 Global IoT Summit (GIoTS)* (2019), pp. 1–6
59. A.J.C. Trappey, C.V. Trappey, U.H. Govindarajan, J.J. Sun, A.C. Chuang, A review of technology standards and patent portfolios for enabling cyber-physical systems in advanced manufacturing. IEEE Access **4**, 7356 (2016)
60. N.K. Nawandar, V.R. Satpute, IoT based low cost and intelligent module for smart irrigation system. Comput. Electron. Agric. **162**, 979 (2019)
61. R. Sharma, A. Shishodia, S. Kamble, A. Gunasekaran, A. Belhadi, Agriculture supply chain risks and covid-19: mitigation strategies and implications for the practitioners. Int. J. Logist. Res. Appl. **0**, 1 (2020)
62. L. Gualtieri, E. Rauch, R. Vidoni, Emerging research fields in safety and ergonomics in industrial collaborative robotics: a systematic literature review. Rob. Comput. Integr. Manuf. **67**, 101998 (2021)
63. M. Fabris, R.M. Abbriano, M. Pernice, D.L. Sutherland, A.S. Commault, C.C. Hall, L. Labeeuw, J.I. McCauley, U. Kuzhiuparambil, P. Ray, T. Kahlke, P.J. Ralph, Emerging technologies in algal biotechnology: toward the establishment of a sustainable, algae-based bioeconomy. Front. Plant Sci. **11**, (2020)
64. K. Vassakis, E. Petrakis, I. Kopanakis, Big Data Analytics: Applications, Prospects and Challenges, in *Mobile Big Data: A Roadmap from Models to Technologies*. ed. by G. Skourletopoulos, G. Mastorakis, C.X. Mavromoustakis, C. Dobre, E. Pallis (Springer International Publishing, Cham, 2018), pp. 3–20
65. R. Nukala, K. Panduru, A. Shields, D. Riordan, P. Doody, J. Walsh, *Internet of Things: A Review from 'Farm to Fork,'* in *2016 27th Irish Signals and Systems Conference (ISSC)* (2016), pp. 1–6

66. A.A. Malik, A. Bilberg, Complexity-based task allocation in human-robot collaborative assembly. Ind. Robot Int. J. Robot. Res. Appl. **46**, 471 (2019)
67. P.C.M. Arachchige, P. Bertok, I. Khalil, D. Liu, S. Camtepe, M. Atiquzzaman, A trustworthy privacy preserving framework for machine learning in industrial IoT systems. IEEE Trans. Ind. Inf. **16**, 6092 (2020)
68. E. Mavridou, E. Vrochidou, G.A. Papakostas, T. Pachidis, V.G. Kaburlasos, Machine vision systems in precision agriculture for crop farming. J. Imaging **5**, 12 (2019)
69. P.W. Khan, Y.-C. Byun, N. Park, IoT-blockchain enabled optimized provenance system for food industry 4.0 using advanced deep learning. Sensors **20**, 10 (2020)
70. L. Trivelli, A. Apicella, F. Chiarello, R. Rana, G. Fantoni, A. Tarabella, From precision agriculture to industry 4.0: unveiling technological connections in the agrifood sector. British Food J. **121**, 1730 (2019)
71. G. Dutta, R. Kumar, R. Sindhwani, R.K. Singh, Digital transformation priorities of india's discrete manufacturing smes–a conceptual study in perspective of industry 4.0. Compet. Rev. Int. Bus. J. **30**, 289 (2020)
72. A. Martín-Gómez, F. Aguayo-González, A. Luque, A holonic framework for managing the sustainable supply chain in emerging economies with smart connected metabolism. Resour. Conserv. Recycl. **141**, 219 (2019)

Chapter 3
The European Union Context on Era 4.0 and Its Dimensions: Relationships with the Agricultural Sector

3.1 Introduction

The digital transformation has held great importance worldwide, namely for the European Union. The reports for the Digital Economy and Society Index are a reflection of the relevance of the digital dimension in the current evolution of economies and societies. This index encompasses areas relating to the following topics [1]: human capital; connectivity; integration of digital dimensions; public services digitalisation; and research and development in ICT. The context of the pandemic accelerated this digital transition and the use of associated technologies [2]. Nonetheless, these transitions to the digital era have brought new challenges and come across, in some circumstances, serious constraints because of the lack of enough human capital to implement the necessary transformations [3]. In these contexts, the focus of the countries involved seems to be different across digitalisation of the business sectors, public services, digital infrastructures and human capital qualification [4].

The digital wave and its related contexts bring about discussion concerning the different associated dimensions, including the statistical assessments and indices considered [5], and comparisons between global contexts [6]. In any case, the digital economy and society index has positive impacts on the labour market, promoting an increase in the rate of employment such as salaries and reducing labour insecurity [7]. Digitalisation appears as one more element of sustainable development [8]. In addition, there are signs of convergence of this index for countries in the European Union across the vast majority of dimensions [9]. The different realities across countries in the European Union [10] have several impacts, including on e-commerce [11]. Some of these differences persist due to specific particularities of each country [12], some of them institutional. The ICT domains hold great importance in promoting, for example, public services for populations [13] through public administration [14] and economic growth [15]. These trends create new opportunities for the labour markets [16]. Nonetheless, this vision of the digital transformation in the European Union based on indicators has generated discussion about the adequacy of this

© The Author(s), under exclusive license to Springer Nature Switzerland AG 2022
V. J. P. D. Martinho, *Trends of the Agricultural Sector in Era 4.0*,
SpringerBriefs in Applied Sciences and Technology,
https://doi.org/10.1007/978-3-030-98959-0_3

approach [17]. For example, the relationships between the level of digitalisation and social dimensions are not linear and are not always positive [18].

Considering this survey, the objective of this study is to analyse the digital economy and society context in countries in the European Union with statistical information obtained from the Eurostat [19] database for the ICT specialists (considering the importance of human capital). Another objective is to assess the relationships of these dimensions with variables related to the socioeconomic frameworks, namely those associated with the agricultural sector (considering data from the FAOSTAT [20]). For these relationships, assessment panel data regressions were considered [21–25]. For data analysis, shapefiles from the Eurostat [19] were considered that were analysed through the QGIS software [26, 27]. For further assessment, spatial autocorrelation analysis was carried out following GeoDa procedures [28, 29]. The results obtained were benchmarked with the findings published by the Digital Economy and Society Index report [30].

3.2 Data Analysis for the Digital Economy and Society

Table 3.1 displays the variables obtained from the Eurostat database which were considered to characterise the qualified human capital related to the digital transition in European Union member states. This table highlights that the enterprises which employed ICT specialists remained stable over the period considered (2014–2020) and around 21%. This stability occurred, in general, for the other variables taken into account. The context of the pandemic, for example, did not bring relevant changes. The percentage of enterprises that had no difficulties in filling vacancies for jobs needing ICT experts is low, showing that there is still some work to be done here, namely in order to increase the availability of specialists for these areas in the labour market. The weight of men in employed ICT specialists is still high. In addition, the majority of the employed ICT experts are between 35 and 74 years of age and have tertiary education (levels 5–8). Finally, the weight of employed ICT experts on the total amount of employment is also almost residual.

Analysing the context across countries in the European Union (on average over the period 2014–2020), Fig. 3.1 shows that Ireland, Belgium, Denmark, Hungary and Netherlands are the countries with a greater percentage of enterprises that employ ICT specialists. Spain, Denmark, Bulgaria, Belgium and Malta are the countries where the greater percentage of enterprises have no difficulties in finding ICT experts (Fig. 3.2). The higher percentages of men as employed ICT specialists are verified in Czechia, Hungary, Slovakia, Malta and Luxembourg (in general, Central Europe, Fig. 3.3) and the greater percentages of women occur in Bulgaria, Romania, Latvia, Lithuania and Finland (Fig. 3.4). The younger employed ICT experts have more expression in Malta, Lithuania, Estonia, Latvia and Romania (Fig. 3.5), and the ICT experts aged 35 and over years are predominantly found in Italy, Denmark, Finland, Sweden and Spain (Fig. 3.6). Italy, Germany, Czechia, Denmark and Malta are the countries with higher percentages for the employed ICT experts having lower

Table 3.1 Percentage of enterprises that employ specialists and percentage of specialists by gender, age, educational level and relative to the total employment, on average over the European Union countries

	2014	2015	2016	2017	2018	2019	2020
Enterprises that employ ICT specialists (%)	21	21	21	21	21	21	21
Enterprise had no hard-to-fill vacancies for jobs requiring ICT specialist skills (%)	5	4	4	4	4	4	4
Employed ICT specialists by gender (% of Males)	83	82	82	82	82	81	81
Employed ICT specialists by gender (% of Females)	17	18	18	18	18	19	19
Employed ICT specialists by age (% of From 15 to 34 years)	41	40	40	40	40	40	39
Employed ICT specialists by age (% of From 35 to 74 years)	59	60	60	60	60	60	61
Employed ICT specialists by educational attainment level (% of Less than primary, primary, secondary and post-secondary non-tertiary education (levels 0–4))	38	36	36	35	34	33	33
Employed ICT specialists by educational attainment level (% of Tertiary education (levels 5–8))	62	63	63	64	66	66	67
Employed ICT specialists—total (% of total employment)	3	4	4	4	4	4	4

educational attainment (Fig. 3.7) and Lithuania, Spain, Cyprus, Ireland and France are those where the employed ICT specialists have a higher level of education are the majority (Fig. 3.8). The higher percentages of employed ICT specialists in total for employment appear in Finland, Sweden, Luxembourg, Estonia and Netherlands (Fig. 3.9).

3.3 Spatial Autocorrelation Analysis

In this section, a spatial autocorrelation analysis will be carried out, namely local spatial autocorrelation, for the variables considered from the digital economy and society, following GeoDa procedures. In figures of this section clusters, high–high and low–low signify positive local autocorrelation (the value of a variable in a country is positively spatially correlated with the values of this variable in neighbouring countries) for high and low values, respectively. The clusters high–low or low–high represent negative spatial autocorrelation [31, 32]. For the spatial analysis, a queen contiguity matrix was considered.

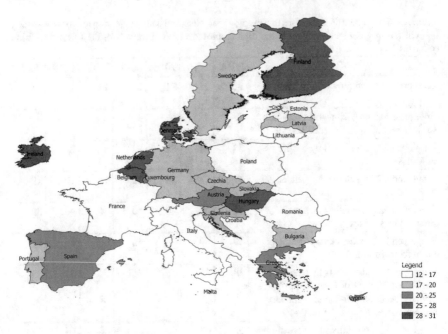

Fig. 3.1 Enterprises that employ ICT specialists (%), on average over the period 2014–2020 and across the European Union countries

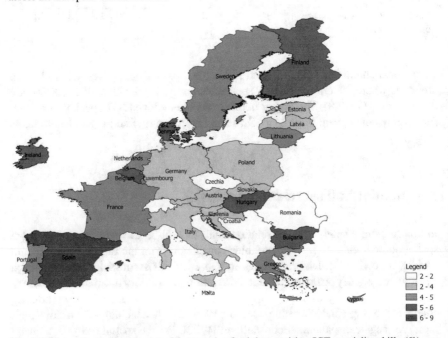

Fig. 3.2 Enterprise had no hard-to-fill vacancies for jobs requiring ICT specialist skills (%), on average over the period 2014–2020 and across the European Union countries

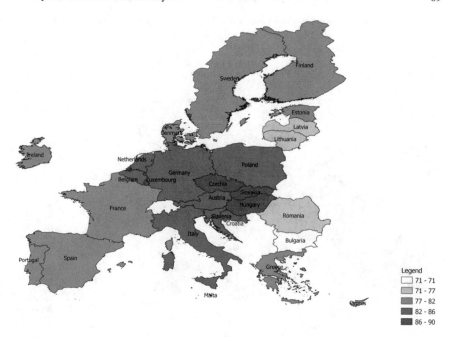

Fig. 3.3 Employed ICT specialists by gender (% of Males), on average over the period 2014–2020 and across the European Union countries

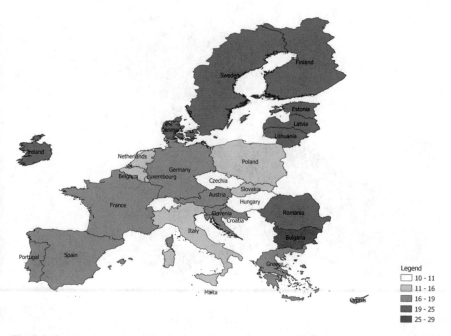

Fig. 3.4 Employed ICT specialists by gender (% of Females), on average over the period 2014–2020 and across the European Union countries

Fig. 3.5 Employed ICT specialists by age (% of From 15 to 34 years), on average over the period 2014–2020 and across the European Union countries

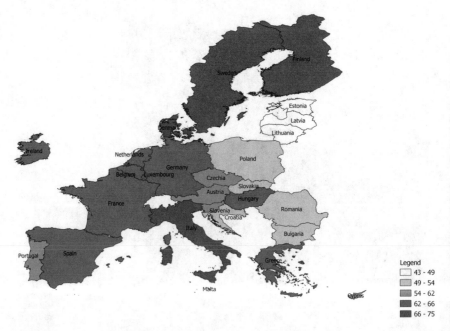

Fig. 3.6 Employed ICT specialists by age (% of From 35 to 74 years), on average over the period 2014–2020 and across the European Union countries

Fig. 3.7 Employed ICT specialists by educational attainment level (% of Less than primary, primary, secondary and post-secondary non-tertiary education (levels 0–4)), on average over the period 2014–2020 and across the European Union countries

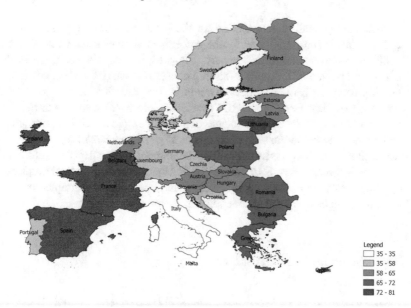

Fig. 3.8 Employed ICT specialists by educational attainment level (% of Tertiary education (levels 5–8)), on average over the period 2014–2020 and across the European Union countries

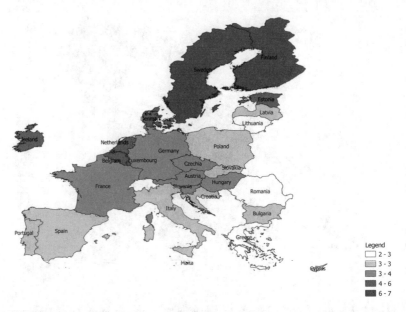

Fig. 3.9 Employed ICT specialists—total (% of total employment), on average over the period 2014–2020 and across the European Union countries

The local spatial autocorrelation for the percentage of enterprises that employ ICT specialists (Fig. 3.10) and for the percentage of enterprises without difficulties in filling their vacancies for ICT experts (Fig. 3.11) is relatively weak. The percentage of male employed ICT experts exhibits a cluster high–high for Germany and neighbouring countries (Fig. 3.12). These countries are parts of a cluster low–low for the percentages of female employed ICT specialists (Fig. 3.13). Baltic countries integrate a cluster high–high for the percentages of younger employed ICT experts (Fig. 3.14) and a cluster low–low for the specialists at 35 or more years of age (Fig. 3.15). For the employed ICT experts by educational attainment, the local spatial autocorrelation is relatively weak (Figs. 3.16 and 3.17). For the percentage of total employed ICT specialists relative to the total for employment, Finland, for example, reveals signs of positive high-high local spatial autocorrelation (Fig. 3.18).

3.4   Panel Data Regressions

Table 3.2 illustrates the results for panel data regressions with spatial effects [23–25, 31, 32], considering the Cobb–Douglas model as a basis [33] linearized (with the variables in logarithms) for the agricultural sector extended with the different percentages for the employed ICT specialists. In this model, the dependent variables are the gross agricultural production value logarithm (constant 2014–2016 I$) and the independent variables are for agricultural employment (employment in crop

Fig. 3.10 Local spatial autocorrelation for enterprises that employ ICT specialists (%), on average over the period 2014–2020 and across the European Union countries

Fig. 3.11 Local spatial autocorrelation for enterprise had no hard-to-fill vacancies for jobs requiring ICT specialist skills (%), on average over the period 2014–2020 and across the European Union countries

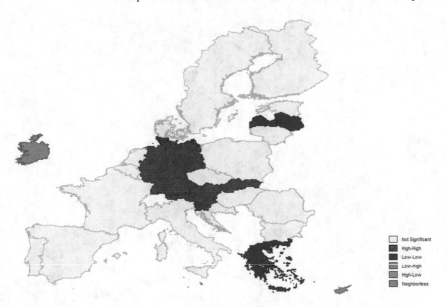

Fig. 3.12 Local spatial autocorrelation for employed ICT specialists by gender (% of Males), on average over the period 2014–2020 and across the European Union countries

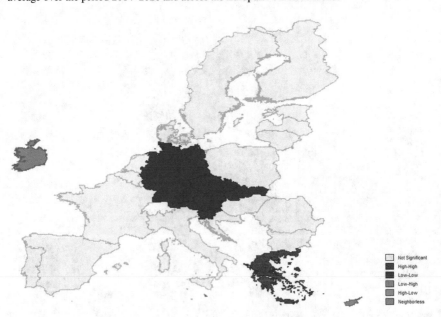

Fig. 3.13 Local spatial autocorrelation for employed ICT specialists by gender (% of Females), on average over the period 2014–2020 and across the European Union countries

Fig. 3.14 Local spatial autocorrelation for employed ICT specialists by age (% of From 15 to 34 years), on average over the period 2014–2020 and across the European Union countries

Fig. 3.15 Local spatial autocorrelation for employed ICT specialists by age (% of From 35 to 74 years), on average over the period 2014–2020 and across the European Union countries

Fig. 3.16 Local spatial autocorrelation for employed ICT specialists by educational attainment level (% of Less than primary, primary, secondary and post-secondary non-tertiary education (levels 0–4)), on average over the period 2014–2020 and across the European Union countries

Fig. 3.17 Local spatial autocorrelation for employed ICT specialists by educational attainment level (% of Tertiary education (levels 5–8)), on average over the period 2014–2020 and across the European Union countries

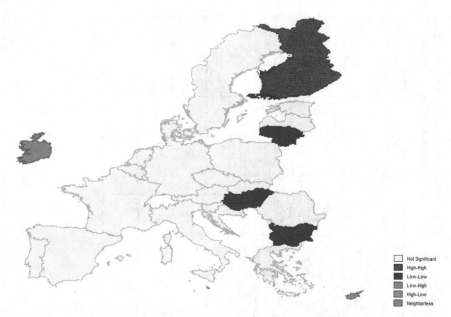

Fig. 3.18 Local spatial autocorrelation for employed ICT specialists—total (% of total employment), on average over the period 2014–2020 and across the European Union countries

and animal production, hunting and related service activities) and the agricultural area. These variables for the agricultural sector were obtained from the FAOSTAT database.

In general, agricultural employment has no statistical significance (when it has it is with negative impacts and low elasticities) and the area always has significance with coefficients between 0.189 and 0.373 (considering fixed effects, given the results for the Hausman test). In the estimations for the employed ICT specialists—total (% of total employment) the Hausman test shows evidence for the random effects and here the coefficient is 0.452.

The coefficients of the variables associated with the percentages of employed ICT experts show positive and statistically significant impacts on the agricultural output for females (although only for random effects) and for younger specialists. Negative and statistically significant effects were found for ICT experts above the age of 35 and for specialists with lower educational attainment. The percentage of ICT experts within the total for employment also has positive and significant impacts.

Table 3.2 Panel data estimations with spatial effects based on the linear Cobb–Douglas model for the agricultural sector and the percentages of employed ICT specialists, over the period 2004–2019 and across the European Union countries, considering the gross agricultural production value logarithm (constant 2014–2016 I$) as dependent variable

Countries	Based on the original model		Based on the original model plus PMSE		Based on the original model plus PFSE		Based on the original model plus PYSE		Based on the original model plus POSE		Based on the original model plus PLLSE		Based on the original model plus PHLSE		Based on the original model plus PTSTE	
Model	Fixed effects	Random effects	Fixed effects	Random effects	Fixed effects	Random effects	Fixed effects	Random effects	Fixed effects	Random effects	Fixed effects	Random effects	Fixed effects	Random effects	Fixed effects	Random effects
Constant		13.660^a (13.500) [0.000]		14.570^a (11.470) [0.000]		14.421^a (12.000) [0.000]		14.923^a (11.250) [0.000]		15.589^a (10.760) [0.000]		13.800^a (12.740) [0.000]		14.601^a (11.570) [0.000]		14.555^a (11.920) [0.000]
Employment logarithm	−0.023 (−1.010) [0.314]	0.003 (0.130) [0.894]	−0.045 (−1.780) [0.075]	−0.010 (−0.360) [0.719]	−0.040 (−1.580) [0.114]	−0.009 (−0.360) [0.718]	−0.051^a (−2.080) [0.037]	−0.015 (−0.570) [0.572]	−0.051^a (−2.090) [0.037]	−0.015 (−0.570) [0.569]	−0.010 (−0.340) [0.736]	0.033 (1.140) [0.254]	−0.060^a (−2.180) [0.029]	−0.019 (−0.660) [0.509]	0.005 (0.200) [0.838]	0.036 (1.330) [0.184]
Area logarithm	0.373^a (5.070) [0.000]	0.555^a (8.350) [0.000]	0.281^a (3.480) [0.001]	0.517^a (6.560) [0.000]	0.283^a (3.320) [0.001]	0.503^a (6.360) [0.000]	0.189^a (2.230) [0.026]	0.458^a (5.070) [0.000]	0.195^a (2.310) [0.021]	0.461^a (5.160) [0.000]	0.312^a (3.830) [0.000]	0.542^a (7.600) [0.000]	0.273^a (3.370) [0.001]	0.516^a (6.440) [0.000]	0.222^a (2.820) [0.005]	0.452^a (5.580) [0.000]
PMSE logarithm			−0.072 (−1.070) [0.285]	−0.052 (−0.750) [0.452]												
PFSE logarithm					0.035 (1.690) [0.090]	0.034^a (2.710) [0.007]										
PYSE logarithm							0.127^a (3.330) [0.001]	0.077 (1.940) [0.052]								
POSE logarithm									−0.166^a (−3.310) [0.001]	−0.104^a (−2.010) [0.044]						

(continued)

Table 3.2 (continued)

Countries	Based on the original model		Based on the original model plus PMSE		Based on the original model plus PFSE		Based on the original model plus PYSE		Based on the original model plus POSE		Based on the original model plus PLLSE		Based on the original model plus PHLSE		Based on the original model plus PTSTE	
PLLSE logarithm											-0.056[a] (-2.120) [0.034]	-0.074[a] (-2.820) [0.005]				
PHLSE logarithm													-0.049 (-1.620) [0.105]	-0.032 (-1.050) [0.293]		
PTSTE logarithm															0.165[a] (5.530) [0.000]	0.160[a] (5.310) [0.000]
Lagged independent variable (Employment logarithm)	-0.253[a] (-4.000) [0.000]	-0.333[a] (-5.590) [0.000]	-0.288[a] (-3.930) [0.000]	-0.346[a] (-5.060) [0.000]	-0.283[a] (-3.900) [0.000]	-0.364[a] (-5.740) [0.000]	-0.292[a] (-4.420) [0.000]	-0.346[a] (-5.520) [0.000]	-0.277[a] (-4.280) [0.000]	-0.338[a] (-5.470) [0.000]	-0.214[a] (-3.110) [0.002]	-0.279[a] (-4.290) [0.000]	-0.284[a] (-4.170) [0.000]	-0.340[a] (-5.360) [0.000]	-0.252[a] (-3.900) [0.000]	-0.318[a] (-5.220) [0.000]
Lagged independent variable (Area logarithm)	0.738[a] (2.880) [0.004]	0.317 (1.820) [0.069]	0.598[a] (2.070) [0.039]	0.226 (1.220) [0.224]	0.645[a] (2.130) [0.033]	0.246 (1.330) [0.182]	0.621[a] (2.180) [0.029]	0.249 (1.330) [0.185]	0.588[a] (2.070) [0.038]	0.233 (1.240) [0.216]	0.651[a] (2.260) [0.024]	0.188 (1.010) [0.314]	0.534 (1.850) [0.064]	0.209 (1.120) [0.263]	0.540 (1.940) [0.052]	0.193 (1.100) [0.273]
Lagged dependent variable	0.167 (1.150) [0.251]	0.000 (0.000) [0.999]	0.185 (1.320) [0.188]	0.069 (0.640) [0.525]	0.208 (1.490) [0.137]	0.066 (0.610) [0.542]	0.188 (1.300) [0.192]	0.059 (0.530) [0.594]	0.196 (1.350) [0.177]	0.065 (0.580) [0.560]	0.202 (1.400) [0.162]	0.055 (0.500) [0.616]	0.191 (1.320) [0.186]	0.077 (0.700) [0.484]	0.200 (1.510) [0.132]	0.078 (0.750) [0.454]
Lagged error term	0.534[a] (4.760) [0.000]	0.614[a] (7.540) [0.000]	0.517[a] (4.510) [0.000]	0.564[a] (6.000) [0.000]	0.491[a] (4.160) [0.000]	0.559[a] (5.960) [0.000]	0.494[a] (4.170) [0.000]	0.556[a] (5.850) [0.000]	0.479[a] (3.930) [0.000]	0.548[a] (5.670) [0.000]	0.474[a] (3.830) [0.000]	0.544[a] (5.620) [0.000]	0.498[a] (4.240) [0.000]	0.550[a] (5.750) [0.000]	0.527[a] (4.910) [0.000]	0.578[a] (6.480) [0.000]

(continued)

Table 3.2 (continued)

Countries	Based on the original model	Based on the original model plus PMSE	Based on the original model plus PFSE	Based on the original model plus PYSE	Based on the original model plus POSE	Based on the original model plus PLLSE	Based on the original model plus PHLSE	Based on the original model plus PTSTE
Hausman test	27.960[a] [0.000]	23.410[a] [0.001]	72.050[a] [0.000]	156.190[a] [0.000]	140.510[a] [0.000]	24.940[a] [0.000]	107.020[a] [0.000]	12.280 [0.091]

[a]Statistically significant at 5%; The variables were lagged with a contiguity matrix; PMSE, Employed ICT specialists by gender (% of Males); PFSE, Employed ICT specialists by gender (% of Females); PYSE, Employed ICT specialists by age (% of From 15 to 34 years); POSE, Employed ICT specialists by age (% of From 35 to 74 years); PLLSE, Employed ICT specialists by educational attainment level (% of Less than primary, primary, secondary and post-secondary non-tertiary education (levels 0-4)); PHLSE, Employed ICT specialists by educational attainment level (% of Tertiary education (levels 5-8)); PTSTE, Employed ICT specialists—total (% of total employment)

The spatial effects are always statistically significant and negative for employment spatially lagged (with a contiguity matrix). This highlights the impacts of agricultural employment of neighbouring countries on the agricultural output of the member states. Impacts from the agricultural area lagged are positive (however, this variable only has statistical significance in some models). The random spatial effects are always positive and with statistical significance.

3.5 Conclusions

The objective of this study was to analyse the digital economy and society context in the European Union, focussing on the information for human capital. The intention was to also analyse the interrelationships between the human capital framework in Era 4.0 and the socioeconomic dynamics of the agricultural sector. For that purpose, statistical information from Eurostat and FAOSTAT was considered. These data were assessed through spatial methodologies and spatial autocorrelation analysis. In order to quantify the different relationships, panel data regressions were also carried out.

The literature review shows the level of attention given by the European Union to digital transformation. A sign of this is the creation of the Digital Economy and Society index. Human capital appears to be crucial for a successful process. The relevance of this digital implementation for sustainability is also highlighted. Nonetheless, there is no consensus among researchers about the contributions of the digital Era to the several dimensions of sustainability, namely for social welfare.

The data analysis highlights that there is still work to be done to improve the percentage of employed ICT specialists in European Union enterprises and to increase the availability of experts in these areas. In addition, there is a need to increase the percentage of women employed as ICT experts, as well as younger and more qualified people. On the other hand, the realties for employed ICT specialists across the various European Union member states is still different. These contexts call for policies that promote a more convergent process in the European digital transition because human capital is decisive for effective digital implementation.

The local spatial autocorrelation is, in general, weak, but there are some clusters of positive spatial correlation that may be considered to implement policies that may be introduced in some strategic countries that could then spread to the neighbouring member states. This may be taken into account in dealing with the low percentage of women employed as ICT experts.

The panel data regressions with spatial effects reveal the importance of both women and younger people as ICT specialists for the dynamics in the European Union agricultural sector, as well as the weight of the total employed ICT experts in the total for employment. In addition, there are statistically significant effects from spatial variables, namely from agricultural employment spatially lagged (effects of the employment from neighbouring countries) and from error term spatially lagged (where the effects from neighbouring countries are random).

In terms of practical recommendation and policy recommendations, it is suggested that European Union institutions joint with each national institution of the member states design strategies to increase human capital with skills adjusted for the digital transition and to promote a convergent process across the various countries in order to avoid contexts emerging at different velocities in these processes. For future research, it could be important to benchmark the European context with other global contexts.

Acknowledgements This work is funded by National Funds through the FCT—Foundation for Science and Technology, I.P., within the scope of the project Refª UIDB/00681/2020. Furthermore we would like to thank the CERNAS Research Centre and the Polytechnic Institute of Viseu for their support.

References

1. G. Grinberga-Zalite, J. Hernik, *Digital Performance Indicators in the Eu*, in *Research for Rural Development 2019*, vol. 2, ed. by S. Treija, S. Skujeniece (Latvia Univ Life Sciences & Technologies, Jelgava, 2019), pp. 183–188
2. M. Stoica, B. Ghilic-Micu, M. Mircea, The telework paradigm in the ioe ecosystem-a model for the teleworker residence choice in context of digital economy and society. Econ. Comput. Econ. Cybern. Stud. **55**, 263 (2021)
3. G. Vitols, I. Arhipova, L. Paura, *Programming Skills Gap Reduction by Extramural School Development: University Success Case Study in Latvia*, in *12th International Technology, Education and Development Conference (Inted)*, ed. by L.G. Chova, A.L. Martinez, I.C. Torres (Iated-Int Assoc Technology Education & Development, Valenica, 2018), pp. 3526–3531
4. N. Volkova, I. Kuzmuk, N. Oliinyk, I. Klymenko, A. Dankanych, development trends of the digital economy: E-Business, E-Commerce. Int. J. Comput. Sci. Netw. Secur. **21**, 186 (2021)
5. Z. Banhidi, I. Dobos, A. Nemeslaki, What the overall digital economy and society index reveals: a statistical analysis of the DESI EU28 dimensions. Reg. Stat. **10**, 42 (2020)
6. Z. Banhidi, I. Dobos, A. Nemeslaki, Comparative analysis of the development of the digital economy in Russia and EU measured with DEA and using dimensions of DESI. Vestn. St. Petersb. Univ. Ekon. **35**, 588 (2019)
7. O. Basol, E.C. Yalcin, How does the digital economy and society index (DESI) affect labor market indicators in EU Countries? Hum. Syst. Manag. **40**, 503 (2021)
8. M. Jovanovic, J. Dlacic, M. Okanovic, Digitalization and society's sustainable development-measures and implications. Zb. Rad. Ekon. Fak. Rijeci **36**, 905 (2018)
9. R. Borowiecki, B. Siuta-Tokarska, J. Maron, M. Suder, A. Thier, K. Zmija, Developing digital economy and society in the light of the issue of digital convergence of the markets in the European Union Countries. Energies **14**, 2717 (2021)
10. M. Jurcevic, L. Lulic, V. Mostarac, The digital transformation of croatian economy compared with Eu member Countries. Ekon. Vjesn. **33**, 151 (2020)
11. B. Jakovic, T. Curlin, I. Miloloza, Enterprise digital divide: website e-commerce functionalities among European Union Enterprises. Bus. Syst. Res. J. **12**, 197 (2021)
12. E. Laitsou, A. Kargas, D. Varoutas, Digital competitiveness in the European Union Era: the greek case. Economies **8**, 85 (2020)
13. M. Decman, The analysis of E-government services adoption and use in slovenian information society between 2014 and 2017. Cent. Eur. Public Adm. **16**, 193 (2018)
14. A.V. Todorut, V. Tselentis, Digital technologies and the modernization of public administration. Qual. Access Success **19**, 73 (2018)

15. A. Fernandez-Portillo, M. Almodovar-Gonzalez, R. Hernandez-Mogollon, Impact of ICT development on economic growth. A study of OECD European Union Countries. Technol. Soc. **63**, 101420 (2020)
16. J. Luis Martinez-Cantos, Digital skills gaps: a pending subject for gender digital inclusion in the European Union. Eur. J. Commun. **32**, 419 (2017)
17. D. Giannone, M. Santaniello, Governance by indicators: the case of the digital agenda for Europe. Info. Commun. Soc. **22**, 1889 (2019)
18. A. Kwilinski, O. Vyshnevskyi, H. Dzwigol, Digitalization of the EU economies and people at risk of poverty or social exclusion. J. Risk Financ. Manag. **13**, 142 (2020)
19. Eurostat, *Several Statistics and Information*. https://ec.europa.eu/eurostat
20. FAOSTAT, *Several Statistics*. https://www.fao.org/faostat/en/#home
21. O. Torres-Reyna, *Panel Data Analysis Fixed and Random Effects Using Stata (v. 4.2)*
22. B.H. Baltagi, *Econometric Analysis of Panel Data*, 4th edn. (Wiley, Chichester, UK ; Hoboken, NJ, 2008)
23. StataCorp, *Stata 15 Base Reference Manual* (Stata Press, College Station, TX, 2017)
24. StataCorp, *Stata Statistical Software: Release 15* (StataCorp LLC, College Station, TX, 2017)
25. Stata, *Stata: Software for Statistics and Data Science*. https://www.stata.com/
26. QGIS, *QGIS Project*. https://www.qgis.org/en/site/
27. QGIS.org, *QGIS Geographic Information System* (QGIS Association, 2022)
28. L. Anselin, I. Syabri, Y. Kho, GeoDa: an introduction to spatial data analysis. Geogr. Anal. **38**, 5 (2006)
29. GeoDa, *GeoDa Software*. https://geodacenter.github.io/
30. European Commission, *DESI\Shaping Europe's Digital Future*. https://digital-strategy.ec.europa.eu/en/policies/desi
31. V.D. Martinho, M.D.C. Sánchez-Carreira, P. Reis Mourão, Transnational economic clusters: the case of the Iberian Peninsula. Reg. Sci. Policy Pract. **13**, 1442 (2021)
32. V.J.P.D. Martinho, Impact of Covid-19 on the convergence of GDP per Capita in OECD Countries. Reg. Sci. Policy Pract. **13**, 55 (2021)
33. C.W. Cobb, P.H. Douglas, A theory of production. Am. Econ. Rev. **18**, 139 (1928)

Chapter 4
Benchmarking the European Union's Digital Context with Those of Other Global Agricultural Frameworks

4.1 Introduction

The digital transition has captured the attention of many different stakeholders, including that of the scientific community. The curiosity concerning the dimension of these topics and the challenges that the digital transformation brings have encouraged the rise of scientific and technical documents and the creation of indices associated with digital implementation. One of these indices is the Digital Adoption Index which assesses digital implementation worldwide considering the dimensions of people (opportunities and welfare), government (efficiency and responsibility of service supply) and business (productivity and growth) [1, 2]. The relevance of this index justifies further studies about their interrelationships [3]. The research already carried out considering this index focus, for example, on the public sector of a selected sample of countries [4] and globalisation [5].

Digital adoption, transformation, implementation and transition are currently common expressions used to characterise this wave associated with Era 4.0 and the adoptions of smart technologies by different economic sectors and public services. Digital adoption in the agricultural sector brings forth new opportunities and challenges, namely in the frameworks of value chains [6]. Other studies have focused on the interrelationships of digital adoption in the following fields: tourism sector [7], regional economies [8], party practices in the UK [9], party organisation [10], firm productivity [11], work councils in Germany [12], populism [13], networking between nongovernmental organisations [14], the health sector [15], disease treatment [16], Malaysian industry [17], Canadian indigenous communities [18], responses to COVID-19 [19], European and Turkish industry [20], digital marketing [21], digital skills [22], older generations [23], higher education institutions [24], e-Government services [25], Portuguese institutions [26], organisations and their employees in the context of the pandemic [27] and labour markets [28].

In fact, this digital wave has interconnections with various dimensions of the economy and society, supporting relevant transformations with positive impacts on

V. J. P. D. Martinho, *Trends of the Agricultural Sector in Era 4.0*,
SpringerBriefs in Applied Sciences and Technology,
https://doi.org/10.1007/978-3-030-98959-0_4

economic dynamics and social well-being. It has also had negative impacts on some social domains and comes with challenges that still require attention, such as those related to women's equal opportunities [29] and benefits [30] in these processes.

Considering the frameworks described, the objective of this research is to assess global digital adoption and its interrelationships with agricultural socioeconomic dynamics. To achieve these aims, statistical information from The World Bank [31] was considered, namely for the Digital Adoption Index and for the socioeconomic farming variables. This information was analysed considering Geographic Information System (GIS) methodologies, factor-cluster analyses, matrices of correlation and cross-section regressions.

4.2   Data Analysis

Figures presented in this section were obtained through QGIS [32], considering shapefiles found in the Eurostat database [33]. The intention was to analyse global digital adoption, taking into account the Digital Adoption Index, and the interrelationships of these contexts with agricultural socioeconomic dynamics (agricultural output and employment). Considering the availability of data, the year 2016 was considered for the analysis carried out.

Figure 4.1 and Table 4.1 for agriculture, forestry and fishing, value added (% of GDP), reveal that Africa and Asia (namely in the South) are countries that maintain a high weight for the agricultural sector within the economy. This is confirmed by

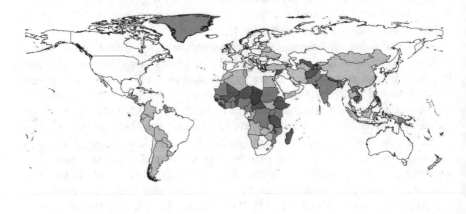

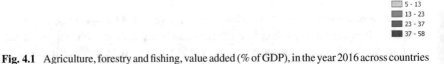

Fig. 4.1 Agriculture, forestry and fishing, value added (% of GDP), in the year 2016 across countries worldwide

Table 4.1 Top 20 countries for agriculture, forestry and fishing, value added (% of GDP), in the year 2016

Countries	Value added
Sierra Leone	58
Guinea-Bissau	46
Chad	46
Mali	37
Liberia	36
Niger	35
Ethiopia	35
Syrian Arab Republic	34
Central African Republic	32
Burundi	32
Comoros	31
Uzbekistan	29
Benin	28
Tanzania	27
Micronesia, Fed. Sts	27
Myanmar	27
Kiribati	26
Malawi	26
Afghanistan	26
Nepal	26

Fig. 4.2 and Table 4.2 for employment in agriculture (% of total employment). For agricultural land (% of land area), the context is a little different and here European and South American countries also appear with more relevance (Fig. 4.3 and Table 4.3). The Digital Adoption index (and the respective sub-indices) seems to follow a different pattern, and in this case, the African countries, in general, have lower scores (Figs. 4.4, 4.5, 4.6 and 4.7 and Tables 4.4, 4.5, 4.6 and 4.7).

4.3 Spatial Autocorrelation Analysis

In this section, local spatial autocorrelation analysis was carried out following GeoDa software [34, 35] procedures. In the figures shown, the clusters high–high and low–low represent positive spatial autocorrelation (the values of a variable in a country are spatially and positively correlated with the values of the same variable in neighbouring countries) for high and low values, respectively. The clusters high–low and low–high are associated with negative local spatial autocorrelation. The clusters for local positive spatial autocorrelation represent countries where for the variable

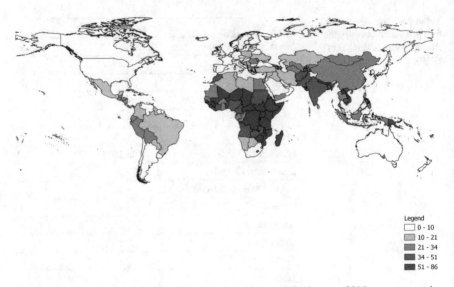

Fig. 4.2 Employment in agriculture (% of total employment) in the year 2016 across countries worldwide

Table 4.2 Top 20 countries for employment in agriculture (% of total employment) in the year 2016

Countries	Employment
Burundi	86
Somalia	81
Malawi	77
Chad	76
Niger	73
Uganda	72
Mozambique	72
Central African Republic	71
Ethiopia	69
Zimbabwe	67
Tanzania	67
Nepal	67
Rwanda	66
Congo, Dem. Rep	66
Madagascar	66
Mali	65
Lao PDR	65
Eritrea	65
Guinea	63
Guinea-Bissau	62

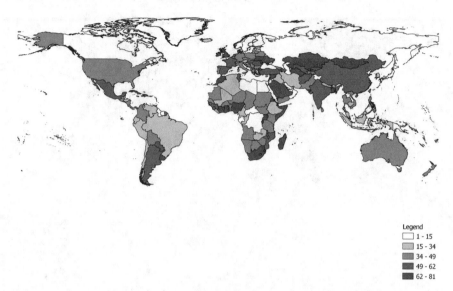

Fig. 4.3 Agricultural land (% of land area) in the year 2016 across countries worldwide

Table 4.3 Top 20 countries for agricultural land (% of land area) in the year 2016

Countries	Area
Saudi Arabia	81
Uruguay	81
Kazakhstan	80
South Africa	79
Burundi	79
Lesotho	78
Syrian Arab Republic	76
Nigeria	76
Eritrea	75
Rwanda	73
Djibouti	73
Mongolia	73
Turkmenistan	72
Uganda	72
United Kingdom	72
Ukraine	72
Isle of Man	71
Eswatini	71
El Salvador	71
Comoros	71

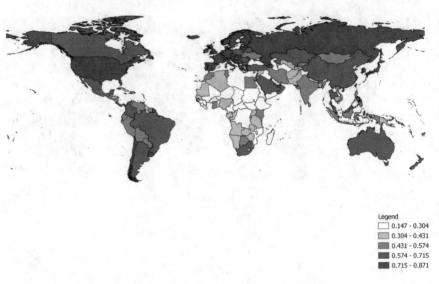

Legend
- 0.147 - 0.304
- 0.304 - 0.431
- 0.431 - 0.574
- 0.574 - 0.715
- 0.715 - 0.871

Fig. 4.4 Digital Adoption index in the year 2016 across countries worldwide

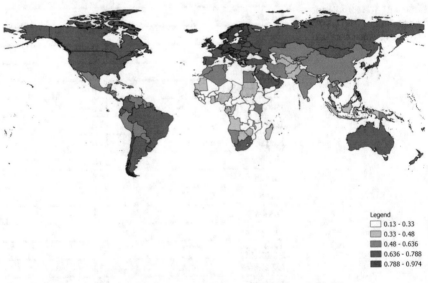

Legend
- 0.13 - 0.33
- 0.33 - 0.48
- 0.48 - 0.636
- 0.636 - 0.788
- 0.788 - 0.974

Fig. 4.5 DAI Business Sub-index in the year 2016 across countries worldwide

considered interventions in one of them may spread to neighbouring countries. This may be important for policy implementation.

For agricultural output and employment weights (Figs. 4.8 and 4.9), there are clusters high–high namely in African countries and low–low in Europe. For the agricultural area weight (Fig. 4.10), there are some clusters high–high in Africa,

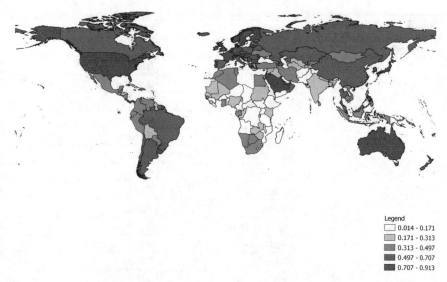

Fig. 4.6 DAI People Sub-index in the year 2016 across countries worldwide

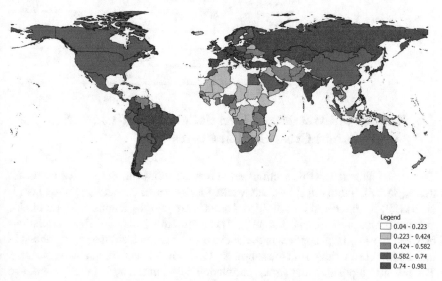

Fig. 4.7 DAI Government Sub-index in the year 2016 across countries worldwide

Europe and Asia. For the Digital Adoption Index, the clusters high–high appear, namely, in South America, Europe and Russia (Fig. 4.11).

Table 4.4 Top 20 countries for Digital Adoption index in the year 2016

Countries	Digital Adoption index
Singapore	0.871
Luxembourg	0.863
Austria	0.862
Korea, Rep	0.858
Malta	0.855
Germany	0.840
Netherlands	0.838
Japan	0.835
Estonia	0.833
Sweden	0.832
United Arab Emirates	0.823
Switzerland	0.822
Finland	0.807
Norway	0.804
Lithuania	0.793
Denmark	0.791
Israel	0.788
Bahrain	0.786
Portugal	0.785
Belgium	0.780

4.4   Results for Matrices of Correlation, Factor-Cluster Analyses and Cross-Section Regressions

The results obtained in this section were obtained following Stata software procedures [36–38]. Table 4.8 shows the results for Spearman´s rank correlation coefficients [39] for the variables considered associated with the Digital Adoption index and the global agricultural variables. There are strong and positive correlations between the Digital Adoption Index and the sub-indices and also among the sub-indices (the correlations with the sub-index DAI Government are relatively weaker). There is also a positive and strong correlation between the agricultural output and employment weights. The same does not happen between these two variables and the agricultural land bias/weight. There are strong correlations, albeit negative, between the agricultural output and employment and the Digital Adoption Index. These contexts are confirmed by Table 4.9 for factor analysis [40–43]. The clusters obtained with the factors found through factor analysis (to avoid problems of collinearity) are those demonstrated in Fig. 4.12. These findings may, for example, provide relevant support for national and international policymakers .

Table 4.5 Top 20 countries for DAI Business Sub-index in the year 2016

Countries	DAI Business Sub-index
Iceland	0.974
Luxembourg	0.944
Malta	0.942
Sweden	0.941
Finland	0.923
Denmark	0.918
Netherlands	0.910
United Kingdom	0.904
Switzerland	0.889
Norway	0.882
Austria	0.877
Germany	0.868
Czech Republic	0.860
Slovenia	0.860
Singapore	0.852
Hong Kong SAR, China	0.851
Belgium	0.850
Estonia	0.847
Barbados	0.833
Andorra	0.829

Considering the linear Cobb–Douglas model [44] as a basis, Table 4.10 shows that, in fact, the agricultural output weight is dependent on agricultural employment (elasticity of 0.675) and is negatively impacted by the Digital Adoption Index (elasticity of −0.515). These results for the Digital Adoption Index are explained by the negative correlation between this index and the global agricultural output. This means that there is still much work to be done in the least developed/developing countries if there is an intention to improve agricultural performance and food security.

4.5   Conclusions

The objective of this study was to assess global digital adoption and to analyse the interrelationships between these frameworks and the agricultural socioeconomic dynamics. Consequently, data from The World Bank for the year 2016 were considered (agriculture, forestry and fishing, value added (% of GDP); employment in agriculture (% of total employment); agricultural land (% of land area); the Digital Adoption Index and sub-indices) that were analysed through GIS approaches, local

Table 4.6 Top 20 countries for DAI People Sub-index in the year 2016

Countries	DAI People Sub-index
Hong Kong SAR, China	0.913
Denmark	0.897
Switzerland	0.890
Luxembourg	0.874
Macao SAR, China	0.869
Austria	0.865
Sweden	0.855
Korea, Rep	0.842
Bahrain	0.840
Japan	0.835
Finland	0.831
Iceland	0.823
Norway	0.811
Singapore	0.803
United Arab Emirates	0.802
Estonia	0.800
United Kingdom	0.799
Netherlands	0.796
New Zealand	0.787
Malta	0.786

spatial autocorrelation analysis, matrices of correlation, factor-cluster analyses and cross-section regressions.

The literature survey highlights that digital adoption is a worldwide reality as it also is for different activities within the economy and society. These transformations bring interesting contributions to the dynamics of enterprises, but with social impacts that need to be addressed further by the various stakeholders.

The data analysis shows that agricultural output and employment still have a great weight within the total economy in African and Asian countries (namely those in the South), but it is in these countries (specifically African) where digital adoption is weaker. To improve food security, international organisations have an important role to play here in order to improve the digital scores of these countries. The local spatial autocorrelation shows that there are some high–high clusters in African countries for the agricultural variables where there may be a consideration to design agricultural policies which could promote the digitalisation of the sector in these nations.

The negative correlations among the agricultural output and employment weights and digital adoption are also highlighted by the matrices of correlation, factor-cluster analyses and cross-section regressions. These contexts need greater attention from international and national institutions, namely in dealing with constraints to implement the digital adoption.

Table 4.7 Top 20 countries for DAI Government Sub-index in the year 2016

Countries	DAI Government Sub-index
Korea, Rep	0.981
Singapore	0.957
Japan	0.909
Chile	0.892
United Arab Emirates	0.886
Uruguay	0.881
Italy	0.873
Portugal	0.871
Germany	0.871
Malaysia	0.867
Estonia	0.853
Israel	0.850
Austria	0.845
Spain	0.840
Kazakhstan	0.839
Malta	0.838
Lithuania	0.827
Russian Federation	0.824
Serbia	0.819
Brazil	0.818

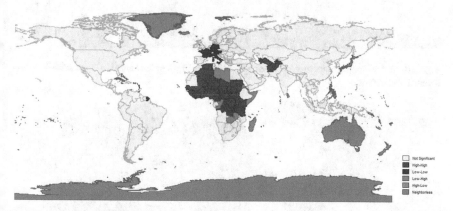

Fig. 4.8 Local spatial autocorrelation for agriculture, forestry and fishing, value added (% of GDP), in the year 2016 across countries worldwide

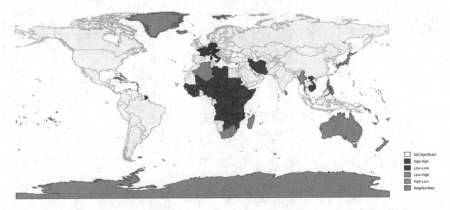

Fig. 4.9 Local spatial autocorrelation for employment in agriculture (% of total employment) in the year 2016 across countries worldwide

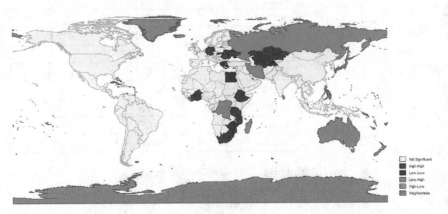

Fig. 4.10 Local spatial autocorrelation for agricultural land (% of land area) in the year 2016 across countries worldwide

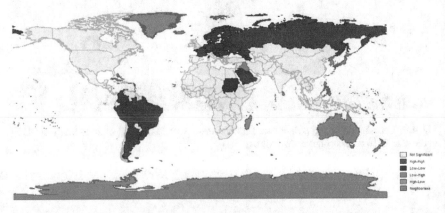

Fig. 4.11 Local spatial autocorrelation for Digital Adoption index in the year 2016 across countries worldwide

Table 4.8 Spearman's rank correlations between the several variables considered for the year 2016 and across countries worldwide

	Agriculture, forestry, and fishing, value added (% of GDP)	Employment in agriculture (% of total employment)	Agricultural land (% of land area)	Digital Adoption index	DAI Business Sub-index	DAI People Sub-index	DAI Government Sub-index
Agriculture, forestry, and fishing, value added (% of GDP)	1.000						
Employment in agriculture (% of total employment)	0.8544[a] (0.000)	1.000					
Agricultural land (% of land area)	0.1812[a] (0.022)	0.1845[a] (0.019)	1.000				
Digital Adoption index	−0.7905[a] (0.000)	−0.8377[a] (0.000)	−0.120 (0.128)	1.000			
DAI Business Sub-index	−0.7885[a] (0.000)	−0.8471[a] (0.000)	−0.123 (0.119)	0.9433[a] (0.000)	1.000		
DAI People Sub-index	−0.8071[a] (0.000)	−0.8734[a] (0.000)	−0.1807[a] (0.022)	0.9546[a] (0.000)	0.9072[a] (0.000)	1.000	
DAI Government Sub-index	−0.5647[a] (0.000)	−0.5605[a] (0.000)	0.019 (0.813)	0.8499[a] (0.000)	0.6902[a] (0.000)	0.7005[a] (0.000)	1.000

[a]Statistically significant at 5%

Table 4.9 Factor analysis for the year 2016 and across countries worldwide

Factor analysis/correlation through principal-component factors and orthogonal varimax

Factor	Variance	Difference	Proportion	Cumulative
Factor1	3.878	2.833	0.646	0.646
Factor2	1.045		0.174	0.820

Rotated factor loadings and unique variances

Variable	Factor1	Factor2	Uniqueness	
Agriculture, forestry, and fishing, value added (% of GDP)	−0.841	0.136	0.274	
Employment in agriculture (% of total employment)	−0.891	0.129	0.191	
Agricultural land (% of land area)	−0.081	0.981	0.031	
DAI Business Sub-index	0.931	−0.067	0.128	

(continued)

Table 4.9 (continued)

Factor analysis/correlation through principal-component factors and orthogonal varimax

Factor	Variance	Difference	Proportion	Cumulative
DAI People Sub-index	0.945	−0.112	0.094	
DAI Government Sub-index	0.781	0.174	0.360	

Kaiser–Meyer–Olkin analysis of sampling adequacy

Variable	kmo			
Agriculture, forestry, and fishing, value added (% of GDP)	0.903			
Employment in agriculture (% of total employment)	0.853			
Agricultural land (% of land area)	0.574			
DAI Business Sub-index	0.852			
DAI People Sub-index	0.791			
DAI Government Sub-index	0.875			
Overall	0.845			

In terms of practical implications and policy recommendations, the suggestion is to reinforce programs for vocational training in digital skills and programs to improve digital infrastructures in the lesser developed countries. Rural and agricultural extension services are also fundamental in modernising the global farming sector. It is here where international organisations are called upon to reinforce their strategies and implement policies of digital transformation in these countries.

For future research, it could be important to analyse the relationships among the agricultural dynamics, digital adoption and sustainability.

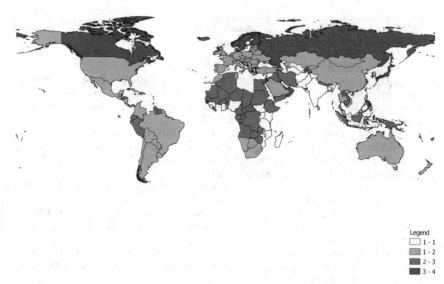

Fig. 4.12 Clusters found considering the factors obtained through factor analysis

Table 4.10 Cross-section regression (agriculture, forestry, and fishing, value added (% of GDP) logarithm, as dependent variable) for the year 2016 and across countries worldwide

	Coefficient	t	P > |t|
Constant	−0.568[a]	−2.530	0.012
Employment in agriculture (% of total employment logarithm	0.675[a]	10.860	0.000
Agricultural land (% of land area) logarithm	0.065	1.020	0.308
Digital Adoption index logarithm	−0.515[a]	−2.760	0.006

[a]Statistically significant at 5%; Ramsey RESET test confirms the adequacy of the model

Acknowledgements This work is funded by National Funds through the FCT—Foundation for Science and Technology, I.P., within the scope of the project Ref UIDB/00681/2020. Furthermore we would like to thank the CERNAS Research Centre and the Polytechnic Institute of Viseu for their support.

References

1. The World Bank, *World Development Report 2016: Digital Dividends*
2. The World Bank, *Digital Adoption Index*. https://www.worldbank.org/en/publication/wdr2016/Digital-Adoption-Index
3. S. Elmassah, E.A. Hassanein, Digitalization and subjective wellbeing in Europe. Digit. Policy Regul. Gov. (n.d.)
4. P. Kokkinakos, O. Markaki, S. Koussouris, J. Psarras, *Digital Transformation: Is Public Sector Following the Enterprise 2.0 Paradigm?*, in *Digital Transformation and Global Society*, ed.

by A.V. Chugunov, R. Bolgov, Y. Kabanov, G. Kampis, M. Wimmer, vol. 674 (Springer International Publishing Ag, Cham, 2016), pp. 96–105

5. M. Skare, D. Riberio Soriano, How globalization is changing digital technology adoption: an international perspective. J. Innov. Knowl. **6**, 222 (2021)

6. H.J. Smidt, Factors affecting digital technology adoption by small-scale farmers in agriculture value chains (AVCs) in South Africa. Inform. Technol. Dev. (n.d.)

7. U. Al-Mulali, S.A. Solarin, A.E. Andargol, H.F. Gholipour, Digital adoption and its impact on tourism arrivals and receipts. Anatolia **32**, 337 (2021)

8. K. Alam, M.O. Erdiaw-Kwasie, M. Shahiduzzaman, B. Ryan, Assessing regional digital competence: digital futures and strategic planning implications. J. Rural Stud. **60**, 60 (2018)

9. K. Dommett, roadblocks to interactive digital adoption? elite perspectives of party practices in the United Kingdom. Party Polit. **26**, 165 (2020)

10. K. Dommett, L. Temple, P. Seyd, Dynamics of intra-party organisation in the digital age: a grassroots analysis of digital adoption. Parliam. Aff. **74**, 378 (2021)

11. P. Gal, G. Nicoletti, C. von Ruden, S. Sorbe, T. Renault, Digitalization and productivity: in search of the holy grail-firm-level empirical evidence from European Countries. Int. Product. Monit. **37**, 39 (2019)

12. S. Genz, L. Bellmann, B. Matthes, Do German works councils counter or foster the implementation of digital technologies? first evidence from the iab-establishment panel. Jahrb. Natl. Okon. Stat. **239**, 523 (2019)

13. D. Guvercin, Digitalization and populism: cross-country evidence. Technol. Soc. **68**, 101802 (2022)

14. N. Hall, H.P. Schmitz, J.M. Dedmon, Transnational advocacy and ngos in the digital era: new forms of networked power. Int. Stud. Q. **64**, 159 (2020)

15. S.E. Kim, A. Logeswaran, S. Kang, N. Stanojcic, L. Wickham, P. Thomas, J.-P.O. Li, Digital transformation in ophthalmic clinical care during the COVID-19 pandemic. Asia Pac. J. Ophthalmol. **10**, 381 (2021)

16. S. Kirrmann, H. Fahrner, T. Bach, M. Hall, S. Glaess, M. Schmucker, E. Gkika, M. Gainey, D. Baltas, F. Heinemann, Digital radiation oncology. Onkologe **24**, 390 (2018)

17. Y.Y. Lee, M. Falahat, B.K. Sia, Drivers of digital adoption: a multiple case analysis among low and high-tech industries in Malaysia. Asia Pac. J. Bus. Adm. **13**, 80 (2021)

18. R. McMahon, Co-developing digital inclusion policy and programming with indigenous partners: interventions from Canada. Inter. Policy Rev. **9**, (2020)ss

19. N.J. Nampoothiri, F. Artuso, Civil society's response to coronavirus disease 2019: patterns from two hundred case studies of emergent agency. J. Creat. Commun. **16**, 203 (2021)

20. G. Nicoletti, C. von Rueden, D. Andrews, Digital technology diffusion: a matter of capabilities, incentives or both?. Eur. Econ. Rev. **128**, 103513 (2020)

21. D. Nunan, M. Di Domenico, Older consumers, digital marketing, and public policy: a review and research agenda. J. Public Policy Mark **38**, 469 (2019)

22. A. Ollerenshaw, J. Corbett, H. Thompson, Increasing the digital literacy skills of regional smes through high-speed broadband access. Small Enterp. Res. **28**, 115 (2021)

23. S. Oppl, C. Stary, Game-playing as an effective learning resource for elderly people: encouraging experiential adoption of touchscreen technologies. Univers. Access Inf. Soc. **19**, 295 (2020)

24. D. Orr, M. Weller, R. Farrow, how is digitalisation affecting the flexibility and openness of higher education provision? results of a global survey using a new conceptual model. J. Interact. Media Educ. **5** (2019)

25. S. Papavasiliou, C. Reaiche, Egovernment digital adoption: can channel choice of individuals be predicted? IADIS Int. J. Comput. Sci. Inf. Syst. **15**, 25 (2020)

26. C.S. Pereira, N. Durao, D. Fonseca, M.J. Ferreira, F. Moreira, An educational approach for present and future of digital transformation in portuguese organizations. Appl. Sci. Basel **10**, 757 (2020)

27. A. Raghavan, M.A. Demircioglu, S. Orazgaliyev, COVID-19 and the new normal of organizations and employees: an overview. Sustainability **13**, 11942 (2021)

28. A.F. Shapiro, F.S. Mandelman, Digital adoption, automation, and labor markets in developing countries. J. Dev. Econ. **151**, 102656 (2021)
29. J. Mariscal, G. Mayne, U. Aneja, A. Sorgner, Bridging the gender digital gap. Economics **13**, 20199 (2019)
30. A.L. Lukyanova, Digitalization and the gender wage gap in Russia. Ekonomicheskaya Polit. **16**, 88 (2021)
31. The World Bank, *Several Statistics and Informations.* https://www.worldbank.org/en/home
32. QGIS.org, *QGIS Geographic Information System* (QGIS Association, 2022)
33. Eurostat, *Several Statistics and Information.* https://ec.europa.eu/eurostat
34. L. Anselin, I. Syabri, Y. Kho, GeoDa: an introduction to spatial data analysis. Geogr. Anal. **38**, 5 (2006)
35. GeoDa, *GeoDa Software.* https://geodacenter.github.io/
36. StataCorp, *Stata 15 Base Reference Manual* (Stata Press, College Station, TX, 2017)
37. StataCorp, *Stata Statistical Software: Release 15* (StataCorp LLC, College Station, TX, 2017)
38. Stata, *Stata: Software for Statistics and Data Science.* https://www.stata.com/
39. C. Spearman, The proof and measurement of association between two things. Am. J. Psychol. **15**, 72 (1904)
40. J.L. Vincent, *Factor Analysis in International Relations: Interpretation, Problem, Areas, and an Application* (Univ Pr of Florida, Gainesville, 1971)
41. J.-O. Kim, C.W. Mueller, *Factor Analysis: Statistical Methods and Practical Issues* (SAGE, 1978)
42. J.-O. Kim, C.W. Mueller, *Introduction to Factor Analysis: What It Is and How To Do It*, 1st edn. (Sage Publications Inc., Beverly Hills, Calif, 1978)
43. O. Torres-Reyna, *Getting Started in Factor Analysis (Using Stata 10) (Ver. 1.5)*
44. C.W. Cobb, P.H. Douglas, A theory of production. Am. Econ. Rev. **18**, 139 (1928)

Chapter 5
Potential Impacts of Era 4.0 on Agricultural Sustainability

5.1 Introduction

Smart farming technologies may provide relevant contributions to increasing the efficiency in the use of the agricultural resources [1] and in this way improve sustainability within the sector [2], through a more adjusted farming management [3]. Sustainability in agriculture is currently of great concern for several stakeholders [4].

In these frameworks, the digital approaches, indeed, allow for new solutions across the whole sector, including irrigation control in hydroponic agriculture [5], agricultural management practices [6] and prediction of internal air temperatures in greenhouses [7]. Among the main added values of these technologies, there is the enormous quantity of information collected, its assessment and its organisation which appear to provide further support to farmers [8].

Big data in the agricultural sector is related to the quantity of information and its variety and facility of access [9]. Information is a crucial factor for production [10] in order to support adjusted decision-making processes [11], allowing for decisions to be made remotely [12]. The management of information from an integrated perspective and in global databases is both the present and the future [13]. Nonetheless, this flow of information brings new concerns about the real use of these enormous quantities of data [14].

In the current contexts of digital agriculture [15] and smart farming [16], new technologies are particularly important in the African context, namely for water management [17], dealing with food insecurity [18] and improving the productivity [19], considering the constraints that will be created by the subsequent conditions associated with global warming [20]. This is true for the Sub-Saharan region and for the Middle East and North Africa (MENA) countries [21]. Africa has a reality with specific characteristics which call for innovative approaches [22]. Public policies [23] and extension services [24] have a relevant role to play here, as well as, those from international organisations.

© The Author(s), under exclusive license to Springer Nature Switzerland AG 2022
V. J. P. D. Martinho, *Trends of the Agricultural Sector in Era 4.0*,
SpringerBriefs in Applied Sciences and Technology,
https://doi.org/10.1007/978-3-030-98959-0_5

The smart farming expression is among the most renowned relating to artificial intelligence in agriculture [25] and is associated with sustainable development [26]. In addition, these approaches, Unmanned Aerial Vehicles and Internet of Things [27], offer new potentialities for farmers worldwide [28], from Asia [29] to Central [30] and North [31] America.

Considering the scenario described earlier, the main aim of this study is to analyse the correlations between global digital adoption and some agricultural variables which may be considered as indicators of the sector's sustainability. For this purpose, data from The World Bank [32] were considered which were analysed through descriptive approaches and matrices of correlation, considering Spearman´s rank correlation coefficients [33] and Stata software procedures [34–36].

5.2   Data Analysis

Figures shown in this section were acquired through data found on The World Bank [32], considering shapefiles obtained from Eurostat [37] and through the QGIS software [38].

Singapore and Luxembourg are the countries with the highest Digital Adoption Index (Fig. 5.1), showing the importance given by these countries to the digital transition. Larger countries, such as China and United States, are those with higher agricultural land space (Fig. 5.2). Burundi and Malawi are where female employment in agriculture (% of female employment) are greater in bias, around 94 and 83%, respectively (Fig. 5.3). The higher values for fertiliser consumption (kilograms per hectare of arable land) appear in Hong Kong and Malaysia (Fig. 5.4), for the food

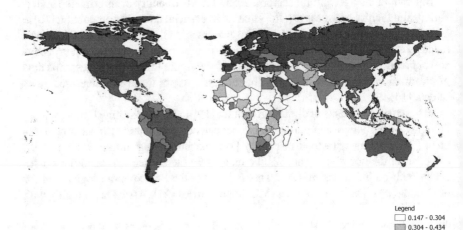

Legend
☐ 0.147 - 0.304
▨ 0.304 - 0.434
▨ 0.434 - 0.574
■ 0.574 - 0.715
■ 0.715 - 0.871

Fig. 5.1 Digital Adoption Index in the year 2016 across countries worldwide

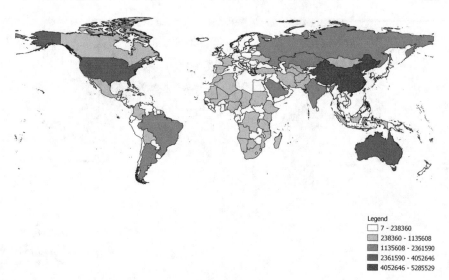

Fig. 5.2 Agricultural land (sq. km) in the year 2016 across countries worldwide

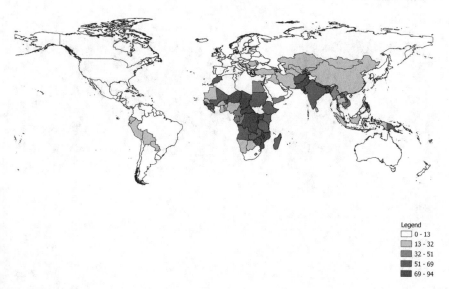

Fig. 5.3 Employment in agriculture, female (% of female employment) in the year 2016 across countries worldwide

production index (2014–2016 = 100) in Grenada and Marshall Islands (Fig. 5.5), for agricultural methane emissions (thousand metric tons of CO_2 equivalent) in India and China (Fig. 5.6), for agricultural methane emissions (thousand metric tons of CO_2 equivalent per sq. km of agricultural land) in the Netherlands and Bangladesh (Fig. 5.7), for agricultural nitrous oxide emissions (thousand metric tons of CO_2

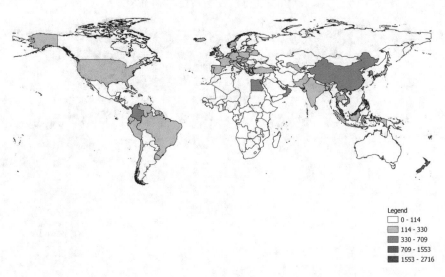

Fig. 5.4 Fertiliser consumption (kilograms per hectare of arable land) in the year 2016 across countries worldwide

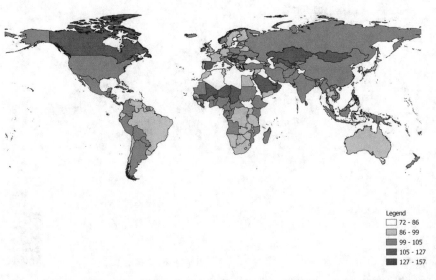

Fig. 5.5 Food production index (2014–2016 = 100) in the year 2016 across countries worldwide

equivalent) in China and India (Fig. 5.8), for agricultural nitrous oxide emissions (thousand metric tons of CO_2 equivalent per sq. km of agricultural land) in Singapore and Brunei (Fig. 5.9), for the rural population in India and China (Fig. 5.10), for the rural population (% of total population) in Burundi and Papua New Guinea (Fig. 5.11) and, finally, for agriculture, forestry and fishing, value added per worker (constant 2015 US$), in Argentina and Iceland (Fig. 5.12).

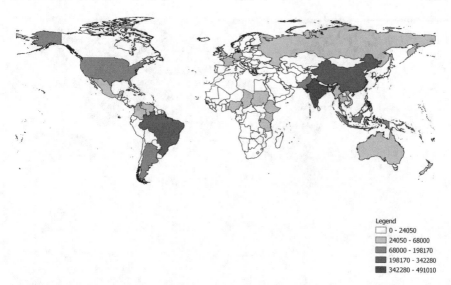

Fig. 5.6 Agricultural methane emissions (thousand metric tons of CO_2 equivalent) in the year 2016 across countries worldwide

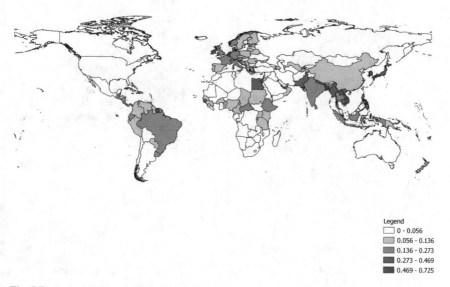

Fig. 5.7 Agricultural methane emissions (thousand metric tons of CO_2 equivalent per sq. km of agricultural land) in the year 2016 across countries worldwide

In general, this data analysis highlights the relevance of the weight of women in agricultural employment and rural population in the total for female employment and total population, respectively, in African countries. In addition, the larger countries, such as China, have the most agricultural land, but also the most total agricultural

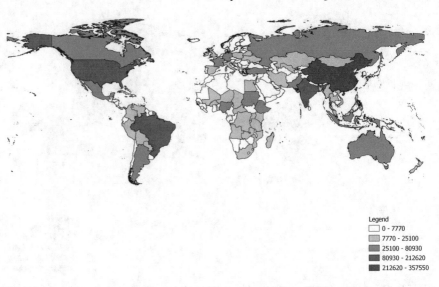

Fig. 5.8 Agricultural nitrous oxide emissions (thousand metric tons of CO_2 equivalent) in the year 2016 across countries worldwide

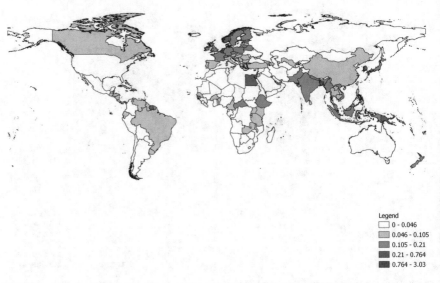

Fig. 5.9 Agricultural nitrous oxide emissions (thousand metric tons of CO_2 equivalent per sq. km of agricultural land) in the year 2016 across countries worldwide

methane and nitrous oxide emissions and total rural population. The values found for agricultural methane and nitrous oxide emissions per sq. Km in countries such as the Netherlands highlight the level of intensification of the farming sector in these nations.

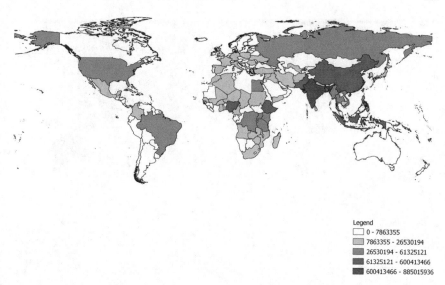

Legend
☐ 0 - 7863355
▨ 7863355 - 26530194
▨ 26530194 - 61325121
▨ 61325121 - 600413466
■ 600413466 - 885015936

Fig. 5.10 Rural population in the year 2016 across countries worldwide

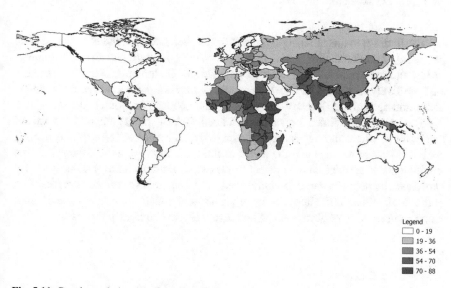

Legend
☐ 0 - 19
▨ 19 - 36
▨ 36 - 54
▨ 54 - 70
■ 70 - 88

Fig. 5.11 Rural population (% of total population) in the year 2016 across countries worldwide

5.3 Results

The results exhibited in Table 5.1 for Spearman´s rank correlation coefficients were obtained following the Stata software procedures.

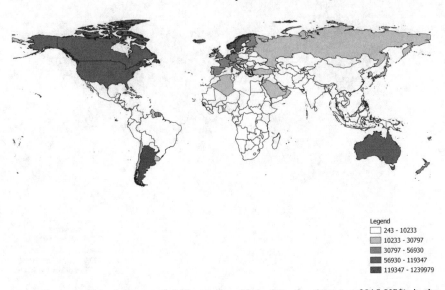

Fig. 5.12 Agriculture, forestry and fishing, value added per worker (constant 2015 US$), in the year 2016 across countries worldwide

The findings in this table highlight that the Digital Adoption Index is strongly and negatively correlated with female employment in agriculture (% of female employment) and rural population (% of total population). On the other hand, it is strongly and positively correlated with fertiliser consumption (kilograms per hectare of arable land) and agriculture, forestry and fishing, value added per worker (constant 2015 US$). Generally, digital adoption is correlated with the most competitive forms of agriculture, where the agricultural productivity (agriculture, forestry and fishing, value added per worker) and the levels of intensification are more expressive. The coefficients of correlation only reveal levels of correlation, and they do not indicate any causality between variables, but it seems that the most developed countries with more competitive agriculture are more prone to adopting these new technologies, improving farming performance, in both cyclical and cumulative processes.

5.4   Conclusions

The main objective of this research was to assess the main interrelationships between global digital adoption and agricultural variables which may be considered as indicators for sustainability in the sector. To achieve these aims, data from The World Bank were considered for the year 2016 and related with the Digital Adoption Index and agricultural indicators related to employment, population, output and environmental impacts. These data were analysed considering GIS (Geographic Information System) approaches and correlations among the variables.

Table 5.1 Spearman's rank correlation coefficients between digital adoption indices and indicators for agricultural sustainability for 2016 and worldwide

	Digital Adoption Index	DAI Business Sub-index	DAI People Sub-index	DAI Government Sub-index	Agricultural land (sq. km)	Employment in agriculture, female (% of female employment)	Fertilizer consumption (kilograms per hectare of arable land)	Food production index (2014–2016 = 100)
Digital Adoption Index	1.000							
DAI Business Sub-index	0.9421[a]	1.000						
	(0.000)							
DAI People Sub-index	0.9526[a]	0.9070[a]	1.000					
	(0.000)	(0.000)						
DAI Government Sub-index	0.8190[a]	0.6466[a]	0.6497[a]	1.000				
	(0.000)	(0.000)	(0.000)					
Agricultural land (sq. km)	−0.2199[a]	−0.2928[a]	−0.2876[a]	0.010	1.000			
	(0.007)	(0.000)	(0.000)	(0.904)				
Employment in agriculture, female (% of female employment)	−0.7593[a]	−0.7824[a]	−0.8022[a]	−0.4431[a]	0.3637[a]	1.000		
	(0.000)	(0.000)	(0.000)	(0.000)	(0.000)			
Fertilizer consumption (kilograms per hectare of arable land)	0.6499[a]	0.6165[a]	0.6326[a]	0.5124[a]	−0.3472[a]	−0.5630[a]	1.000	
	(0.000)	(0.000)	(0.000)	(0.000)	(0.000)	(0.000)		
Food production index (2014–2016 = 100)	−0.034	−0.074	−0.053	0.029	0.031	0.035	−0.045	1.000
	(0.682)	(0.373)	(0.522)	(0.724)	(0.711)	(0.672)	(0.589)	
Agricultural methane emissions (thousand metric tons of CO_2 equivalent)	−0.104	−0.2006[a]	−0.1823[a]	0.130	0.8590[a]	0.2994[a]	−0.118	0.010
	(0.207)	(0.014)	(0.026)	(0.114)	(0.000)	(0.000)	(0.152)	(0.900)
Agricultural methane emissions (thousand metric tons of CO_2 equivalent per sq. km of agricultural land)	0.1952[a]	0.160	0.1976[a]	0.1668[a]	−0.3510[a]	−0.1797[a]	0.4367[a]	−0.082
	(0.017)	(0.052)	(0.016)	(0.042)	(0.000)	(0.028)	(0.000)	(0.321)

(continued)

Table 5.1 (continued)

	Digital Adoption Index	DAI Business Sub-index	DAI People Sub-index	DAI Government Sub-index	Agricultural land (sq. km)	Employment in agriculture, female (% of female employment)	Fertilizer consumption (kilograms per hectare of arable land)	Food production index (2014–2016 = 100)
Agricultural nitrous oxide emissions (thousand metric tons of CO_2 equivalent)	-0.069	-0.152	-0.156	0.148	0.8816[a]	0.2802[a]	-0.117	-0.009
	(0.402)	(0.064)	(0.058)	(0.071)	(0.000)	(0.001)	(0.157)	(0.912)
Agricultural nitrous oxide emissions (thousand metric tons of CO_2 equivalent per sq. km of agricultural land)	0.3840[a]	0.3749[a]	0.3682[a]	0.2927[a]	-0.5244[a]	-0.3434[a]	0.6149[a]	-0.101
	(0.000)	(0.000)	(0.000)	(0.000)	(0.000)	(0.000)	(0.000)	(0.222)
Rural population (% of total population)	-0.7349[a]	-0.7058[a]	-0.7501[a]	-0.5089[a]	0.113	0.7562[a]	-0.5033[a]	0.025
	(0.000)	(0.000)	(0.000)	(0.000)	(0.172)	(0.000)	(0.000)	(0.762)
Rural population	-0.3366[a]	-0.4429[a]	-0.4192[a]	-0.002	0.7684[a]	0.5358[a]	-0.2670[a]	0.027
	(0.000)	(0.000)	(0.000)	(0.982)	(0.000)	(0.000)	(0.001)	(0.745)
Agriculture, forestry, and fishing, value added per worker (constant 2015 US$)	0.8302[a]	0.8451[a]	0.8540[a]	0.5228[a]	-0.2794[a]	-0.8956[a]	0.5679[a]	-0.033
	(0.000)	(0.000)	(0.000)	(0.000)	(0.001)	(0.000)	(0.000)	(0.689)

	Agricultural methane emissions (thousand metric tons of CO_2 equivalent)	Agricultural nitrous oxide emissions (thousand metric tons of CO_2 equivalent)	Agricultural nitrous oxide emissions (thousand metric tons of CO_2 equivalent per sq. km of agricultural land)	Rural population (% of total population)	Rural population	Agriculture, forestry, and fishing, value added per worker (constant 2015 US$)
Agricultural methane emissions (thousand metric tons of CO_2 equivalent per sq. km of agricultural land)						
Digital Adoption Index						
DAI Business Sub-index						
DAI People Sub-index						

(continued)

Table 5.1 (continued)

	Agricultural methane emissions (thousand metric tons of CO$_2$ equivalent)	Agricultural methane emissions (thousand metric tons of CO$_2$ equivalent per sq. km of agricultural land)	Agricultural nitrous oxide emissions (thousand metric tons of CO$_2$ equivalent)	Agricultural nitrous oxide emissions (thousand metric tons of CO$_2$ equivalent per sq. km of agricultural land)	Rural population (% of total population)	Rural population	Agriculture, forestry, and fishing, value added per worker (constant 2015 US$)
DAI Government Sub-index							
Agricultural land (sq. km)							
Employment in agriculture, female (% of female employment)							
Fertilizer consumption (kilograms per hectare of arable land)							
Food production index (2014–2016 = 100)							
Agricultural methane emissions (thousand metric tons of CO$_2$ equivalent)	1.000						

(continued)

Table 5.1 (continued)

	Agricultural methane emissions (thousand metric tons of CO$_2$ equivalent)	Agricultural methane emissions (thousand metric tons of CO$_2$ equivalent per sq. km of agricultural land)	Agricultural nitrous oxide emissions (thousand metric tons of CO$_2$ equivalent)	Agricultural nitrous oxide emissions (thousand metric tons of CO$_2$ equivalent per sq. km of agricultural land)	Rural population (% of total population)	Rural population	Agriculture, forestry, and fishing, value added per worker (constant 2015 US$)
Agricultural methane emissions (thousand metric tons of CO$_2$ equivalent)	0.106 (0.200)	1.000					
Agricultural nitrous oxide emissions (thousand metric tons of CO$_2$ equivalent)	0.9609[a] (0.000)	−0.008 (0.923)	1.000				
Agricultural nitrous oxide emissions (thousand metric tons of CO$_2$ equivalent per sq. km of agricultural land)	−0.159 (0.053)	0.7583[a] (0.000)	−0.134 (0.104)	1.000			
Rural population (% of total population)	0.102 (0.214)	−0.014 (0.863)	0.060 (0.464)	−0.1668[a] (0.042)	1.000		

(continued)

Table 5.1 (continued)

	Agricultural methane emissions (thousand metric tons of CO$_2$ equivalent)	Agricultural methane emissions (thousand metric tons of CO$_2$ equivalent per sq. km of agricultural land)	Agricultural nitrous oxide emissions (thousand metric tons of CO$_2$ equivalent)	Agricultural nitrous oxide emissions (thousand metric tons of CO$_2$ equivalent per sq. km of agricultural land)	Rural population (% of total population)	Rural population	Agriculture, forestry, and fishing, value added per worker (constant 2015 US$)
Rural population	0.8128[a]	−0.037	0.8114[a]	−0.2011[a]	0.4124[a]	1.000	
	(0.000)	(0.655)	(0.000)	(0.014)	(0.000)		
Agriculture, forestry, and fishing, value added per worker (constant 2015 US$)	−0.1969[a]	0.153	−0.156	0.3376[a]	−0.7561[a]	−0.4666[a]	1.000
	(0.016)	(0.063)	(0.058)	(0.000)	(0.000)	(0.000)	

[a]Statistically significant at 5%

The literature review highlights the relevance of smart farming technologies in improving productivity and efficiency in the agricultural sector and from this perspective as determinant approaches towards improving sustainability in farms. In these frameworks, Big Data appears as an interesting solution for collecting, processing and providing information in large quantities, with ease and at high speed. These methodologies may provide substantial hope in dealing with the problems associated with food security in African countries, for example, however, the constraints are still numerous, showing that there is still much work to be done.

The data analysis shows that, in fact, there are problems with agricultural performance in many worldwide countries where solutions are often not easy to find and where there are problems of agricultural intensification having a serious environmental impact, on other parts of the world, which still remains to be solved after many attempts. It seems that in practice it will not be easy to address these challenges even with the new technologies corresponding to digital transition.

Generally, the results obtained with the matrices of correlation confirm these concerns. In fact, the countries with a greater Digital Adoption Index are also those with higher productivity and greater fertiliser consumption. In summary, it would seem that digital transition is interrelated with economic dimensions, but is still less so with environmental and social dimensions.

In terms of practical implication and policy recommendations, the suggestion would be to design strategies that better interlink digital transformations with environmental and social dimensions in the agricultural sector, for a more sustainable development. In addition, for digital adoption to be a reality in the lesser developed/developing countries, a clear and effective approach will be needed from the international community in conjunction with national authorities and populations. For future research, bringing more insight into an effective digital implementation and in a more globally balanced way is suggested.

Acknowledgements This work is funded by National Funds through the FCT—Foundation for Science and Technology, I.P., within the scope of the project Ref$^{\underline{a}}$ UIDB/00681/2020. Furthermore we would like to thank the CERNAS Research Centre and the Polytechnic Institute of Viseu for their support.

References

1. V. Moysiadis, P. Sarigiannidis, V. Vitsas, A. Khelifi, Smart farming in Europe. Comput. Sci. Rev. **39**, 100345 (2021)
2. A.T. Balafoutis, F.K. Van Evert, S. Fountas, Smart farming technology trends: economic and environmental effects, labor impact, and adoption readiness. Agronomy Basel **10**, 743 (2020)
3. A. Belanche, A. Ignacio Martin-Garcia, J. Fernandez-Alvarez, J. Pleguezuelos, A.R. Mantecon, D.R. Yanez-Ruiz, Optimizing management of dairy goat farms through individual animal data interpretation: a case study of smart farming in Spain. Agric. Syst. **173**, 27 (2019)
4. M.E. Latino, A. Corallo, M. Menegoli, B. Nuzzo, Agriculture 4.0 as enabler of sustainable agri-food: a proposed taxonomy. IEEE Trans. Eng. Manage. (n.d.)

5. C. Cambra, S. Sendra, J. Lloret, R. Lacuesta, Smart system for bicarbonate control in irrigation for hydroponic precision farming. Sensors **18**, 1333 (2018)
6. K.-Y. Li, N.G. Burnside, R.S. de Lima, M.V. Pecina, K. Sepp, V.H. Cabral Pinheiro, B.R.C.A. de Lima, M.-D. Yang, A. Vain, K. Sepp, An automated machine learning framework in unmanned aircraft systems: new insights into agricultural management practices recognition approaches. Remote Sens. **13**, 3190 (2021)
7. G. Codeluppi, L. Davoli, G. Ferrari, Forecasting air temperature on edge devices with embedded AI. Sensors **21**, 3973 (2021)
8. G. Kakamoukas, P. Sarigiannidis, G. Livanos, M. Zervakis, D. Ramnalis, V. Polychronos, T. Karamitsou, A. Folinas, N. Tsitsiokas, *A Multi-Collective, IoT-Enabled, Adaptive Smart Farming Architecture*, in *2019 Ieee International Conference on Imaging Systems & Techniques (Ist 2019)* (IEEE, New York, 2019)
9. A. Lytos, T. Lagkas, P. Sarigiannidis, M. Zervakis, G. Livanos, Towards smart farming: systems, frameworks and exploitation of multiple sources. Comput. Netw. **172**, 107147 (2020)
10. C. Eastwood, M. Ayre, R. Nettle, and B. Dela Rue, Making sense in the cloud: farm advisory services in a smart farming future, NJAS-Wagen. J. Life Sci. **90–91**, 100298 (2019)
11. V. Saiz-Rubio, F. Rovira-Mas, From smart farming towards agriculture 5.0: a review on crop data management. Agronomy Basel **10**, 207 (2020)
12. C. Verdouw, B. Tekinerdogan, A. Beulens, S. Wolfert, Digital twins in smart farming. Agric. Syst. **189**, 103046 (2021)
13. I.W. Widayat, M. Koppen, *Blockchain Simulation Environment on Multi-Image Encryption for Smart Farming Application*, in *Advances in Intelligent Networking and Collaborative Systems (Incos-2021)*, ed. by L. Barolli, H.C. Chen, H. Miwa, vol. 312 (Springer International Publishing Ag, Cham, 2022), pp. 316–326
14. E. Jakku, B. Taylor, A. Fleming, C. Mason, S. Fielke, C. Sounness, P. Thorburn, If they don't tell us what they do with it, why would we trust them? trust, transparency and benefit-sharing in smart farming, NJAS-Wagen. J. Life Sci. **90–91**, 100285 (2019)
15. M.K. Sott, L.S. da Nascimento, C.R. Foguesatto, L.B. Furstenau, K. Faccin, P.A. Zawislak, B. Mellado, J.D. Kong, N.L. Bragazzi, A bibliometric network analysis of recent publications on digital agriculture to depict strategic themes and evolution structure. Sensors **21**, 7889 (2021)
16. R. Tombe, *Computer Vision for Smart Farming and Sustainable Agriculture*, in *2020 Ist-Africa Conference (Ist-Africa)*, ed. by M. Cunningham, P. Cunningham (IEEE, New York, 2020)
17. P. Abbott, A. Checco, D. Polese, *Smart Farming in Sub-Saharan Africa: Challenges and Opportunities*, in *Proceedings of the 10th International Conference on Sensor Networks (Sensornets)*, ed. by R.V. Prasad, N. Ansari, C. BenaventePeces (Scitepress, Setubal, 2021), pp. 159–164
18. L. Doyle, L. Oliver, C. Whitworth, *Design of a Climate Smart Farming System in East Africa*, in *2018 IEEE Global Humanitarian Technology Conference (Ghtc)* (IEEE, New York, 2018)
19. S.O. Olawuyi, A. Mushunje, Heterogeneous treatment effect estimation of participation in collective actions and adoption of climate-smart farming technologies in South-West Nigeria. Geo J. **85**, 1309 (2020)
20. J.A. Strauss, P.A. Swanepoel, M.C. Laker, H.J. Smith, Conservation agriculture in rainfed annual crop production in South Africa. S. Afr. J. Plant Soil **38**, 217 (2021)
21. R.A. Bahn, A.A.K. Yehya, R. Zurayk, Digitalization for sustainable agri-food systems: potential status, and risks for the MENA region. Sustainability **13**, 3223 (2021)
22. L. Boronyak, B. Jacobs, A. Wallach, J. McManus, S. Stone, S. Stevenson, B. Smuts, H. Zaranek, Pathways towards coexistence with large carnivores in production systems. Agric. Human Values (n.d.)
23. G. Branca, A. Arslan, A. Paolantonio, U. Grewer, A. Cattaneo, R. Cavatassi, L. Lipper, J. Hillier, S. Vetter, Assessing the economic and mitigation benefits of climate-smart agriculture and its implications for political economy: a case study in Southern Africa. J. Clean Prod. **285**, 125161 (2021)
24. M. Chitakira, N.Z.P. Ngcobo, Uptake of climate smart agriculture in peri-urban areas of South Africa's economic hub requires up-scaling. Front. Sustain. Food Syst. **5**, 706738 (2021)

25. J.P. Garcia Vazquez, R. Salomon Torres, D.B. Perez Perez, Scientometric analysis of the application of artificial intelligence in agriculture. J. Scientometr. Res. **10**, 55 (2021)
26. J.A. Turner, L. Klerkx, T. White, T. Nelson, J. Everett-Hincks, A. Mackay, N. Botha, Unpacking systemic innovation capacity as strategic ambidexterity: how projects dynamically configure capabilities for agricultural innovation. Land Use Pol. **68**, 503 (2017)
27. N. Islam, M.M. Rashid, F. Pasandideh, B. Ray, S. Moore, R. Kadel, A review of applications and communication technologies for internet of things (IoT) and unmanned aerial vehicle (UAV) based sustainable smart farming. Sustainability **13**, 1821 (2021)
28. A. Walter, R. Finger, R. Huber, N. Buchmann, Smart farming is key to developing sustainable agriculture. Proc. Natl. Acad. Sci. USA **114**, 6148 (2017)
29. S.F.P.D. Musa, K.H. Basir, Smart farming: towards a sustainable agri-food system. Br. Food J. **123**, 3085 (2021)
30. E. Collado, A. Fossatti, Y. Saez, Smart farming: a potential solution towards a modern and sustainable agriculture in panama. Aims Agric. Food **4**, 266 (2019)
31. M.T. Saez Parra, R. Ferraz-Almeida, tools and techniques to mitigate communications failures in IoT projects (internet of things) in area with smart irrigation in "sustainable farming. Rev. Gest. Tecnol. **20**, 287 (2020)
32. The World Bank, *Several Statistics*. https://www.worldbank.org/en/home
33. C. Spearman, The proof and measurement of association between two things. Am. J. Psychol. **15**, 72 (1904)
34. StataCorp, *Stata 15 Base Reference Manual* (Stata Press, College Station, TX, 2017)
35. StataCorp, *Stata Statistical Software: Release 15* (StataCorp LLC, College Station, TX, 2017)
36. Stata, *Stata: Software for Statistics and Data Science*. https://www.stata.com/
37. Eurostat, *Several Statistics and Information*. https://ec.europa.eu/eurostat
38. QGIS.org, *QGIS Geographic Information System* (QGIS Association, 2022)

Printed in the United States
by Baker & Taylor Publisher Services